HOW ENLIGHTENMENT
CHANGES YOUR BRAIN

HOW ENLIGHTENMENT
CHANGES YOUR BRAIN

○

The New Science
of Transformation

Andrew Newberg, MD, and Mark Robert Waldman

Avery • an imprint of Penguin Random House • New York

an imprint of Penguin Random House LLC
375 Hudson Street
New York, New York 10014

Most Avery books are available at special quantity discounts for bulk purchase
for sales promotions, premiums, fund-raising, and educational needs. Special
books or book excerpts also can be created to fit specific needs. For details,
write SpecialMarkets@penguinrandomhouse.com.

Library of Congress Cataloging-in-Publication Data

Names: Newberg, Andrew B., 1966- | Waldman, Mark Robert.
Title: How enlightenment changes your brain : the new science of
transformation / Andrew Newberg, MD, and Mark Robert Waldman.
Description: New York City : Avery, 2016.
Identifiers: LCCN 2015025731 | ISBN 9781594633454
Subjects: LCSH: Spirituality. | Awareness. | Insight. |
Enlightenment—Miscellanea. | Neurosciences—Religious aspects.
Classification: LCC BL624 .N485 2016 | DDC 204/.2—dc23
LC record available at http://lccn.loc.gov/2015025731

Printed in the United States of America
1 3 5 7 9 10 8 6 4 2

Book design by Lauren Kolm

CONTENTS

PART 3   Moving Toward Enlightenment

ACKNOWLEDGMENTS

Every book involves dozens of people to bring one's vision to fruition, and Mark and I would like to express our deepest appreciation to everyone who has worked with us over the past two decades. I especially want to thank the thousands of anonymous contributors who have shared their spiritual experiences with us through our surveys and brain-scan studies.

I would like to acknowledge my close colleagues that I have worked with over the years. In particular, Dr. Daniel Monti has become a great friend and colleague as the director of the Myrna Brind Center of Integrative Medicine at Thomas Jefferson University. He has been incredibly supportive of all of my work. My two wonderful mentors, Dr. Abass Alavi and the late Eugene d'Aquili, allowed me to explore this fascinating intersection of the brain and spirituality, always encouraging me to tread into uncharted waters. And Nancy Wintering has been a steadfast collaborator on all of these exciting projects.

We extend our gratitude to Chris Manning, PhD, at Loyola Marymount University, Los Angeles, for helping us to clarify our Spectrum of Human Awareness. We also thank Yuval Ron, a scholar of the Abrahamic mystical traditions and their music, for his input and guidance regarding the complexities of Sufi practices and beliefs.

Our deepest appreciation goes to our agent, Jim Levine, and to our beloved editor, Caroline Sutton, who has brilliantly guided us through our last three books. Also, our heartfelt thanks goes to Brittney Ross, our coeditor for this project, and to Brianna Flaherty, our wonderful copy editor. We send an especially big hug to Bo Rinaldi, who gave us the inspiration to frame this book around the topic of personal enlightenment.

And of course, we could not do this work without the support of our wonderful families, particularly our wives, Stephanie and Susan. While enlightenment is always deeply personal, the experience becomes lacking if we cannot share it with those we love and work with every day.

AUTHORS' NOTE

For over a decade, Mark and I have worked together exploring the nature of consciousness, spirituality, and the brain. For this book, since we have used much of my own research to describe enlightenment, we will generally be using "I" to refer to myself (Andrew), unless otherwise indicated.

But since Mark and I work closely in building our models and hypotheses, I will often use "we" to reflect our collaborative efforts. In addition, research is never a solitary venture, so you'll often find references to "our" work, which also includes the members of my research staff and colleagues with whom I have worked for many years.

We have attempted to make the information in this book as "user friendly" as possible. But generalizations often leave out important subtleties and complexities. So for those who are interested in exploring these topics in more depth, we have provided extensive peer-reviewed references to substantiate the conclusions we have reached.

PART 1

THE ROOTS OF ENLIGHTENMENT

●

I awake

Like bursting ice

In a water jar. [1]

—Basho, seventeenth-century Zen poet

ONE

The Enlightenment of a Troubled Kid

Have you ever had an experience that completely and wholly changed your life? An experience that changed the way you thought *and* the way you acted? An experience that entirely changed your views about your job, your relationships, and your life in general? Many people have. For some, it converted their religious or spiritual beliefs. For others, it may have convinced them that God doesn't exist. It's the type of experience that can totally change the direction of your life, leading you down new paths of discovery.

Perhaps such an experience hasn't happened to you, but you know that you are *looking* for something that will profoundly change your life and provide you with a new sense of meaning and purpose. People have questions, really *big* questions, and they are seeking answers, really *big* answers. They'll spend a fortune on self-help books and courses that promise to transform them, only to be disappointed.

Still, most of us continue to struggle to find "it." But almost everyone has gotten a glimmer of those big, life-changing experiences. Even the smallest moments of clarity harbor some of the same basic elements of the "it" we feel driven to find.

The "it"—the transformation we seek—is what most people refer to as Enlightenment, with a big "E." Eastern philosophy makes a big deal out of big "E" Enlightenment, but in the West, philosophers talk about another form of enlightenment, a small "e" mini-experience that provides us with new insights about ourselves and the world. Throughout this book, we will distinguish the big "E" experiences by using the capital letter for "Enlightenment," and when we refer to the smaller moments, we will use the lowercase "enlightenment." These smaller experiences—these little "e" enlightenments—are great to have and are very helpful for understanding the big "E" Enlightenment. In fact, our research shows that the smaller experiences might even prime our brains to have those grand life-changing transformations. The big Enlightenment experiences are the ones that ultimately relieve suffering and bring peace and happiness to people. And that is the type of experience that the human brain appears to crave.

We want to show you what big "E" Enlightenment is all about— how it affects your life and how it affects your body and brain— and we are going to use three tools to help enlighten you about Enlightenment and then guide you through specific exercises that can help you find it for yourself. The first tool involves the stories of people who have had big "E" and little "e" experiences. These stories are mostly from our online survey, which collected personal descriptions of over two thousand spiritual experiences. We'll share with you some of the remarkable discoveries we've gleaned from these amazing encounters with Enlightenment and what we've learned about how one's beliefs can either promote or inhibit our ability to transform our lives and our brain.

The second tool is a new model of human awareness, a "spectrum" that begins with instinctual awareness and ends with the experience of Enlightenment. As we progress along this spectrum, we are actually moving from a minimal amount of awareness about the world toward a complete awareness of the whole universe. This map combines ancient wisdom and modern science in a way that makes it easy to identify where you are on your path and quest for Enlightenment.

The third tool we'll use is the series of brain-scan studies we've conducted on people who engage in very powerful and unusual forms of spiritual practice involving healing, chanting, channeling, and radical forms of meditation that profoundly alter the normal functioning of the brain. We believe that these studies can offer insights into a faster way to experience the big "E" forms of Enlightenment that are often described in ancient spiritual texts.

MY JOURNEY INTO UNCERTAINTY

I've been mapping the neural correlates of spiritual experiences for nearly three decades, and many people ask me about how I got involved in a field fraught with peril for any aspiring scientist. My career has had its challenges, but the rewards have been phenomenal, and my work continues to reflect my passion for understanding how we, as human beings, grapple with reality as we try to make sense of our world.

So let me share with you how my own journey began and one of the transformational experiences I've had that reshaped my way of thinking about everything. I will try to describe it the best I can, but to this day I struggle with explaining what I experienced. After all, any level of "enlightenment" is almost impossible to relate in words. So as you listen to my story, it's important to keep this in mind: enlightenment, large or small, is an indescribable experience that alters the brain and our awareness of ourselves and the world in a way we find deeply meaningful. And think about your own life-changing experiences, now and throughout the book, to help you find the meaning in your own life.

Growing up, I was a troubled kid, but not in the usual sense of the word. I actually had a wonderful childhood. I had a close relationship with my parents and I got most of the things that I wanted and needed. I was a very happy boy.

Except for one thing: I could never understand why so many people had different beliefs. Why were there so many religions, so many political systems, and so many different views on what was right and wrong? And why did everyone feel so strongly about their beliefs, to the point of inflicting violence on one another? In short, I wanted to get to what was real so I could *know* the truth and not just believe. I would argue that this was my first conscious decision to seek enlightenment, to begin a path that would help illuminate the questions that were burning in my mind. This, by the way, is the dictionary definition of small "e" enlightenment: *to shed light upon a topic of inquiry.*

Unfortunately, my questioning did not lead me to answers; instead, it took me into deeper realms of confusion. This existential uncertainty stayed with me throughout high school and into my college years, and when I tried to talk to my family and friends about these matters, they usually gave me quizzical looks. Some of my teachers even told me I was wasting my time thinking about such questions, but I couldn't let it go. Instead, it became my personal mission to unravel these mysteries of the mind.

I pored over the philosophies of great historical figures, paying particular attention to how they grappled with the nature of reality. I also read many of the world's sacred texts—the Bible, the Quran, the Bhagavad Gita—anything I could find in the library. I read Aristotle, Aquinas, Hume, and Husserl, and I talked to rabbis, priests, and the occasional Buddhist master. The Eastern philosophers gave me insights into the big "E" forms of transformation and the Western philosophers highlighted the "aha" moments of insight that fueled their passion to understand the world rationally. Again, I would call those the little "e" moments, but historians called the apex of Western philosophy the Age of Enlightenment.

In the end, none of these great exemplars of knowledge brought me peace of mind. So I turned my attention to science to see what it had to say about the fundamental laws of reality. I looked into evolution, DNA, cosmology, and neuroscience, but even in these "sacred" halls of academia I never seemed to find the answer to my query. To me, each of these wisdom schools was simply a different system of many beliefs—beliefs that were created and processed by the human brain, a wonderful but faulty device.

Even the most rigorous scientific studies seemed flawed or incomplete at best, with each new piece of research offering contradictory advice. Science is an excellent way to observe the world around us, but it never answered my fundamental question: What is the *real* reality, and why does everyone experience it in a different way? But I still thought that the more I studied the brain, the more I could unravel some of the bigger mysteries of life.

So entering medical school was a particularly exciting time, and I began to explore the brain and body in greater detail than I ever had before. I eventually decided to take an extra year while in medical school in order to study the brain in more detail, and I was introduced to the relatively new technologies related to brain imaging. Now I could begin to see what was going on in a living brain as a person performed different activities or reflected on different ideas and beliefs. For me, this was one of the most exciting experiences in my life. Perhaps now I could find a way to link my pursuit of the deep questions with how my brain was actually trying to answer them.

Then one day, as part of a summer internship, I volunteered to undergo an fMRI scan while performing various memory tasks. After about an hour in the scanner, with my head strapped to the table inside the giant magnetic donut, my back was in agony, my arms felt numb, and I really had to go to the bathroom. I answered all of the memory questions as best I could, but I realized that the researchers were never going to know what was really happening in my mind. All they knew were my answers to the memory tasks. They thought I was simply remembering the different words being presented to me. They had no idea about all of the other things I was thinking and feeling.

At that moment, I had what I'd call a small epiphany, a little "e" enlightenment: *no one can ever know for certain what is going on in another person's mind and brain.* This discovery is now supported by hundreds of studies, but it also relates to another great conundrum for all of cognitive neuroscience. I realized that we can never even fully know what is going on inside our own mind because there are just too many variables involved. In any given minute of our own cognitive awareness, we can have hundreds— maybe thousands—of discrete thoughts, feelings, and sensations constantly flowing in and out of our consciousness.

This insight helped me realize even more how difficult my pursuit of truth and reality was going to be. I concluded once and for all that I had to stop relying on what others had to say about truth, and I also concluded that science was going to leave me somewhat short of my goal. After all, it was my brain that was interpreting whatever information science provided me.

So instead of seeking wisdom through scientific studies, reading books, and talking to other people, I turned my pursuit inward. I reasoned, perhaps naively, that if the best scientists, philosophers, and theologians couldn't agree on these fundamental issues, maybe the answer could be found *within* me. It seemed to me that if I am part of reality, I should be able to quiet down all of my rushing thoughts and try to identify those absolute truths. After all, that's what the quest for Enlightenment is all about, according to those ancient Chinese teachers who had promised me that there was an answer to everything.

By turning inward, I quickly realized the next problem. My mind seemed filled with so many feelings, thoughts, and beliefs, how was I to know which ones could anchor me in reality? As a neuroscientist, I ended up exploring this issue with my research collaborator, Mark Waldman, in our book *Born to Believe*. We documented how the brain can build wildly inaccurate but useful maps about ourselves, the world, and the reality that exists outside our inner perceptions. We *think* we see the world correctly, but we aren't aware of how distorted these maps can be.

As I reflected on the problem of how my own brain—my own mind—was trying to find truth, I found myself becoming more contemplative. I wasn't doing a formal practice like Transcendental Meditation or Vipassana, just my own concoction of thinking about things in a different way, looking for that nugget of truth I could rely on.

At first, I thought this would help get me closer to my goal of understanding reality, but I didn't seem to get any closer. Eventually my agitations returned and I began to question my previous insights.

This, by the way, turns out to be a common experience for enlightenment seekers: we have these moments of insights, these "aha" experiences, and we think we've discovered a fundamental truth. We feel uplifted and incredibly blissful, for a moment, and then our old reality—our familiar habitual mind-set and beliefs—returns. Those who meditate regularly often have little "e" moments of enlightenment, but then the teacher comes by and says "this too will pass," a gentle reminder that the student has yet to experience that big "E" moment where one's entire worldview permanently changes, providing a totally new sense of meaning and answering those big questions.

To have an insight, only to realize that it isn't as helpful as you thought, is one of the most stressful experiences one can have, and that is where I found myself. I started to doubt every thought and belief I had. I didn't feel I knew the truth about anything, and everything seemed more like an opinion and not a fact. I felt trapped in a realm of perpetual doubt, but I had no choice: I had to continue my contemplative search for some fundamental truth. It was a lonely process, occasionally interrupted when I came across someone else who had gone down a similar path, like René Descartes, one of the most important Age of Enlightenment philosophers of the seventeenth century.

I was drawn to his *Meditations on the First Philosophy*, in large part because it seemed to incorporate the two things I felt I was doing—meditating and philosophizing. I became more excited when I read his opening comment: "Of the things which may be brought within the sphere of the doubtful."[1] "Ah," I thought, "he's talking about me!" He is struggling with the same doubt that I am! As I read on, I found some comfort in his famous conclusion *"Cogito ergo sum"*: "I think therefore I am." But then I began to doubt that as well. How did he *know* that there was an

"I" doing the thinking? I felt that there was still something I was missing.

THE INFINITE SEA OF DOUBT . . . AND BLISS

I decided to try a new contemplative experiment. Since I believed that all my thoughts were nothing more than products of an imaginative brain, I attempted to exclude all of the products of my mind: language, feelings, perceptions, self-reflections—anything that could be biased or distorted. My realm of doubt, and the pile of concepts included in that doubt, just grew and grew. I kept forcing myself to find something that was beyond doubt, but I couldn't.

As my doubt escalated, I realized that my entire strategy had to be doubted. Again, this idea was met with great internal pain. How could I know if what I was doing was the right thing? I had to doubt my whole process of doubting—a weird notion, but one that I felt compelled to do.

At that moment, I heard a tiny inner voice whispering, "Stop trying." It reminded me of something my Hindu philosophy professor in college had said when he was discussing how seekers in the Hindu tradition had reached the level of big "E" Enlightenment, the state in which we suddenly rise above our own individual ideas about the world and arrive at a totally new understanding of our self and the universe. My teacher said that finding Enlightenment—catching that ineffable glimmer of a transcendental truth—required a combination of striving and *not* striving. Instead of trying to *find* the "answer," let the answer *come* to you.

Now I want to point out that I was not consciously seeking enlightenment, big or small, I was simply trying to discover a basic truth. But I was failing, so I decided to trust that inner voice to see what "answer" might come to me. I just waited. But for what? I spent two years doing this and it was probably the most psychologically painful period in my life.

But then it happened. I was doing one of my daily philosophical meditations when, suddenly, instead of finding doubt in everything, everything literally became the doubt, with a big "D."*

I found myself floating in what I can only describe as a sea of Infinite Doubt, and it was the most intense, compelling experience I had ever had. Although twenty-five years have passed since that remarkable day, I still find it hard to put that experience into words. The "Infinite Sea" was everywhere, and everything was wrapped up in it—the world, religion, science, philosophy, even my own self! All I had ever wanted to do was eradicate doubt and I ended up finding out that the only certainty is Doubt. All I could do was to surrender myself to it and fully accept and immerse myself in that Infinite Sea in which everything—me, my thoughts, other people, the universe—was unified and connected.

This, as we will explain throughout the book, is one of the com-

---

*In case you're wondering about my use of capital letters, it turns out, from our most recent research, that many extraordinary experiences defy language. Thus I, like many other people, will use a capital letter to signify how big or intense the experience was. The capitalization "deifies" the word and people will often use this tool to differentiate between inconsequential "gods" and the "God" they consider to be supreme. Throughout this book, we'll use this same strategy to address the small insights—the mini-enlightenments—that appear to be essential as one travels down the path toward an ultimate Truth or Enlightenment.

mon elements found in the experiences people refer to as Enlightenment: feeling a sense of unity and surrender coupled with a feeling of profound clarity that some deeper insight or truth or wisdom had been reached. In my case, the experience happened as the result of many years of personal contemplation, but as I would later learn by studying other people's descriptions and conducting numerous brain-scan experiments on various spiritual practices, there are many paths that can produce similar experiences.

A few years ago, I shared my personal experience with my coauthor Mark. When I got to the description of Infinite Doubt, he posed one of the most interesting questions I had ever been asked: "Wasn't that experience terrifying? I would imagine it would be for me."

I paused for a second, and realized that it was actually the most comfortable and blissful experience I had ever had. I understood Mark's confusion; after all, I myself hated the years I had spent questioning and doubting everything. And now that same doubt was everywhere I looked. But perhaps he couldn't realize my bliss because he hadn't *experienced* it for himself. For me, it felt like the weight of the world had been permanently lifted from my shoulders. How strange to feel that! Where had all of my troubles and worries gone? Somehow the doubt was no longer something to be feared but rather something to be embraced. That was the key. Instead of fighting the doubt, I became united *with* it. And that Infinite Sea had it all. It was incredibly intense, profoundly clear, entirely uplifting, deeply emotional, and extraordinarily pleasurable. In fact, it became the most important turning point in my life and philosophy. I felt transformed and the changes that took place inside me twenty-five years ago are still alive today.

For one, I became keenly aware that everybody's beliefs were tenuous, shaped by the creative imagination of the brain. Suddenly, all beliefs were equal, and none were better or worse, or more right or wrong, than the others. They were all partial notions—just glimmers of a reality that may or may not exist beyond the limited perceptions of our mind. I felt a tremendous sense of humility at our inability to really know what is going on in the universe around us. I also felt an intense sense of awe knowing that in spite of our incredibly imperfect brains, we all have an intuitive way of carving a meaningful path through this world.

Since that moment, I have often returned to my transformational experience as a reminder that we are significantly blinded by our beliefs about the world. These limited beliefs are often the cause of our failings, fears, and sufferings because we think we know something when we don't. It's probably one of the most difficult things to realize how trivial and petty these limited beliefs can be.

My experience transformed my life in another way by opening the door to the line of research to which I've devoted the last two decades of my life: identifying the neurological pathways of spirituality and consciousness that can lead us toward those moments of insight and bliss. My goal in writing this book is to share with you the newest research demonstrating that the personal transformation that comes with enlightenment experiences is not just a possibility—it's a biological imperative that drives us from the moment we are born. We also want to offer you some shortcuts that may speed up your own quest for a small "e" or big "E" experience.

WHY ENLIGHTENMENT?

So what can we call my own personal experience? It was powerful, and it changed my life. But I began to wonder if other people's life-transforming experiences were all the same or different from one another. In fact, here in the West, most people don't talk about Enlightenment very much. We all seek happiness and success, friendship and intimacy. But as my career progressed, I felt that while people want to change their life in dramatic ways, there is a great fear about really shaking up their fundamental beliefs about the world. We enjoy the simple insights and "aha" moments—the little "e" experiences—but we rarely dare to rattle our habitual ways of thinking and behaving. We explored this in our book *Born to Believe* by showing how entrenched we are in our beliefs. It is hard to break out of them even when we really want to.

Think about it for a moment: Do you really want to change your life in a radical and profound way or are you just seeking to improve bits and pieces of your life? The point is that the big "E" Enlightenment should not only be for the rare devoted monk or saint but for you as well. And if our data have anything to say about it, we are all capable of reaching those truly big experiences.

For some people, Enlightenment is the most incredible, powerful, and life-transforming experience a person can ever have. At its most basic level, it sheds new light and knowledge on how you think everything in the world is supposed to be. For some, this can be a deeply religious or spiritual experience. For others, it can be incredibly rational. And from our perspective, whatever the person feels, it is also felt deep within the workings of the human brain.

While others have written about Enlightenment from a purely spiritual or philosophical perspective, we are going to talk about it from both a neuroscientific perspective and a personal one. We are going to describe what happens in the brain as people work and move toward enlightenment, what happens at those peak moments of life-changing power, and how the brain is permanently changed as the result of these experiences. And we will reveal the powerful descriptions of the experiences of our survey participants and how the critical elements of enlightenment are reflected in different brain processes. We've used our brain-scan studies to develop a series of exercises that will show you how to facilitate your own personal growth and transformation.

The literature shows that theoretically everyone is capable of experiencing Enlightenment, but our research shows that for those who do have big "E" and little "e" insights, no two people will ever experience them exactly the same way. In other words, it's a highly personal event. That's why we've included many stories gathered from people who participated in a web-based survey I have been running since 2008. People not only wrote about their experiences, they also provided us with data about their personal history and belief systems.

For those who do experience Enlightenment, we discovered

that they—like me—often struggled to find the right words to capture the event in a meaningful way. Here is a sampling of how conscious they were of this problem:

> I can only tell you the general parts of the experience because some details are impossible to describe.
>
> I was totally filled with an essence of love, but I can't find the right words to describe it. It felt like the air around us was made out of love.
>
> It is impossible for me to communicate with you—because this communication is false, and I am false for trying to communicate it.

On and on, our survey participants tried to describe the indescribable, and they always seemed to feel a bit of remorse when they failed to find the right words. But each of them knew what the experience was in their own mind and heart, and each of them came to see the world in a new way, one that guided them toward a newer and deeper sense of reality.

WIRED FOR "E"

The ability to experience enlightenment, big or small, appears to be "wired" into our brain and consciousness. If we can learn to access this function, I believe that everyone would find immense benefits, not just to one's self, but to society as well.

Neurologically speaking, small enlightenment experiences appear to be associated with the most recently evolved structures in our brain, structures that help us find meaning and purpose in our lives. These same neurological circuits help us to regulate our emotions and to generate empathy and compassion toward others. In other words, *neurological* enlightenment—and in particular, our ability to observe ourselves as being *separate* from our daily thoughts and feelings—improves our inner state of well-being *and* our ability to cooperate with others without conflict. Small enlightenment experiences appear to be essential for improving our relationships with family, friends, and colleagues, and we believe that the conscious *search* for the big "E" experience increases our ability to alleviate not only our own suffering but the suffering of others in the world.

There is growing scientific evidence that brief moments of enlightenment occur in most of us, and that the more we consciously seek out these "aha" moments that allow us to see greater truths about ourselves and others, the more likely we are to have a life-changing big "E" experience. Our research, along with the work of some of our colleagues, has identified many qualities of the small "e" experience. For example:

- It can instantly illuminate a difficult problem.
- It immediately interrupts worries, fears, and doubts.
- You'll often feel deeper kindness, compassion, and empathy.

- You'll become more open-minded and tolerant of others.
- You'll feel a deep sense of peace.

These are qualities often described by our survey participants, and they are also qualities that many people discover through meditation and prayer. But the experiences are often transient, whereas Enlightenment brings an instantaneous and permanent change to one's personality or worldview—they actually go from the *experience* of Enlightenment to the *state* of Enlightenment. We will discuss the state of Enlightenment later in the book, but it is important to realize that the state of Enlightenment represents all of the transforming changes brought about by the experience. The Enlightened person has a new sense of meaning and purpose in his life, feels differently about his job and relationships, and no longer fears failure or even death. And we have documented such changes in our survey.

But here's the problem. We do not have scientific "proof" that big "E" experiences represent reality. We have thousands of years of anecdotal stories, but firsthand descriptions are often dismissed in science because we don't know if the person made it up. Has anyone ever captured or measured a big "E" event? We're not sure. But we believe that our most recent brain-scan studies have uncovered a very unusual neurological change that occurs when people say they are experiencing something that resembles Enlightenment. We also see long-term structural changes in the brain in people who are consciously seeking Enlightenment. For these reasons, the evidence suggests that the path *toward* Enlightenment is not only real, but that we are biologically predisposed to seek it. Whether or not we *achieve* it is another matter. Science can neither prove nor disprove God's existence or nonexistence, and the same is true for Enlightenment. If the idea and the experience of Enlightenment feel meaningful and valuable, then by all means pursue it!

Our brain-scan studies also show that there are specific techniques that anyone can use to help speed up the enlightenment process by priming the brain for such experiences. For example, you can learn how to alter your "everyday" consciousness that keeps you pinned to your old beliefs. There are also imagination exercises that you can use to deliberately enter a unique state of consciousness that taps into the creative centers of the brain. You can practice specific sound and movement rituals that will alter your perception of the world. You can also train your brain to form new neural connections that will allow you to feel better about yourself and more tolerant of others. These enlightenment experiences—be they large or small—may even help a person to overcome deep-rooted personality problems and addictions. Simply put, if you seek enlightenment, you will discover a different "you," one who makes life more meaningful and rich.

In Part 1 of this book, we'll reveal the biological, psychological, and cultural roots of the human quest for Enlightenment, describing many personal examples. Some, like Tolstoy's experience, are famous, but the most important examples described in this book come from the two thousand people who voluntarily participated in our online spirituality survey. Some of these individuals experienced Enlightenment spontaneously, while others found it through deep reflection and contemplation. Many were religious, but many were not, and we'll show you the commonalities of these experiences that led us to our formula for helping elicit big "E" and small "e" events within the brain. And we'll introduce you to our Spectrum of Human Awareness to show how the brain can move through different small "e" states on its way to Enlightenment.

We'll also look at the insights gleaned by those who sought Enlightenment through drugs and even by those who found their Enlightenment just doing everyday things. These personal stories, combined with our latest brain imaging studies, will give you, the reader, a road map to create your own path of transformation—and by transformation, we mean a change so significant that you experience yourself as a different person, free from the inner suffering that so often fills one's life.

In Part 2, we'll show you what happens in the brain during a variety of powerful spiritual practices that people have used to induce different states of consciousness as part of their own journey to find enlightenment. These different states move you along a broad spectrum of awareness and can yield some of the small "e" enlightenment experiences as well as help prime your brain for the big "E" Enlightenment experiences.

In Part 3, we'll guide you through a series of concentration and meditation exercises with the goal of helping to prime your brain for Enlightenment. First, we'll teach you some of the most effective ways to relax and observe the wanderings of your busy brain, steps that will prepare your brain to enter creative realms where enlightenment experiences begin to form. Then we'll introduce you to a series of intense practices designed to radically change your brain's activity. Your inner and outer reality will abruptly change, and when this happens, it becomes possible— or so we believe—to experience profound changes in your entire worldview.

PEERING INTO MY "ENLIGHTENED" BRAIN

Many years after my initial experience with Infinite Doubt, when I was fully engaged in doing brain-scan research on spiritual practices, psychological health, and neurological disorders, I began to experiment with the powerful technology of functional magnetic resonance imaging, or fMRI. These machines have been the heart and soul of modern brain research and can track moment-to-moment changes in brain activity that occur when a person performs any specific task.

I wondered if my personal experience of Infinite Doubt—which I can easily trigger by going into a contemplative state—would show up as a specific pattern of brain activity. So I went into the fMRI scanner, and with the magnetic coils surrounding my head and all the banging noises it makes, I proceeded to contemplate Infinite Doubt. I did not tell my assistants exactly what I would be doing other than that it was my own brand of meditation.

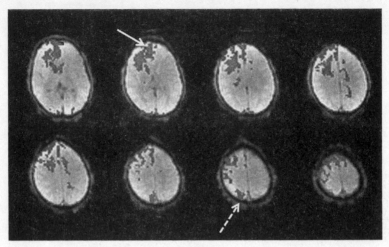

Figure 1.

The imaging results were quite amazing (see Figure 1), and suggested to me that there are specific areas of the brain that can be associated with Enlightenment experiences. My scans showed decreased activity in the parietal lobe (dotted arrow) in the back of the brain, which is consistent with what we see in other people when they feel an intense sense of unity. This has always made sense to me since the usual function of the parietal lobe is to take all of the sensory information coming into the brain and help us create our sense of self and establish how that self is related to the rest of the world—what I refer to as the self-other dichotomy. So a decrease of activity in this area would be associated with a loss of the sense of self and a blurring of the lines between my self and the rest of the world. I would feel deeply connected to the entire world since there are no longer any clear boundaries between it and myself.

Surprisingly, my scans also showed *decreased* frontal lobe activity (solid arrows). Usually we see *increased* frontal lobe activity when people purposefully concentrate on a sacred object, an image, or a specific prayer. I wondered if this unusual decrease might be essential when it comes to mystical or Enlightenment experiences. As each of the studies in this book exemplifies, it turns out that my hunch may be correct: Enlightenment appears to involve a rapid and radical decrease in frontal lobe activity, something that a person can consciously manipulate with his thoughts, intentions, and the use of movement, sound, and breathing.

ENLIGHTENMENT IS "REAL"

Based on our scientific evidence, I now believe that the stories found in sacred texts describing Enlightenment are real in that they are related to specific neurological events that can permanently change the structure and functioning of the brain. People may or may not actually be connecting to God or the supernatural, but ultimately there is something very powerful going on *inside* the brain.

Enlightenment appears to involve a sudden shift of consciousness that temporarily interrupts the way the brain normally responds to the world. These experiences can spontaneously occur or they can suddenly erupt after years of contemplative practice. They can also be triggered by a dramatic or traumatic event, but there appears to be an underlying common thread: Enlightenment can be induced by radically altering the blood flow in different parts of the brain. When this happens, you'll see the world in a new way, often with an incredible sense of awe. But remember: this sudden perceptual shift is a highly *subjective* experience, one that is difficult to describe in words.

From the survey results I've gathered, from the people I've scanned, and from my own personal encounters with this mysterious realm of consciousness, I have come to realize how powerful and positive these experiences are. No wonder people from every time and culture have sought these experiences out. I hope from the stories and research we are about to present that you will be encouraged to seek and find your own path toward Enlightenment.

TWO

What Is Enlightenment?

Knowledge—Gnosis—Wisdom—Science—Reason—Oneness—Unity

Ecstasy—Awakening—Bliss—Purity—Liberation—Insight—Truth

Transcendence—Transformation—Self-Realization—Illumination

Clarity—Inner Peace—Holiness—Revelation—God

Emptiness—Selflessness—Pure Consciousness

Throughout history, different scholars have used every one of these terms to define and capture the essence of Enlightenment. But since the experience is profoundly individual and often difficult to describe in words, it leads many people to different conclusions. For example, the great philosopher Plato offered an elegant metaphor for both big "E" and small "e" enlightenment in his "Allegory of the Cave," in which a group of people have been imprisoned since childhood. They are chained in such a way that they can only see the wall, and behind them a fire burns, casting shadows of themselves on the walls of the cave. The prisoners are fascinated by these shadows and wonder what they are about. As time passes, they build all kinds of beliefs about the shadows and they assume those beliefs are true.

One day, one of the prisoners breaks free, and he turns around to see the fire and the other people who are casting the shadows. Initially he is shocked and surprised, then intrigued, realizing that things are very different from what he believed. This represents the little "e" enlightenment, the initial "aha" experience that begins to change his worldview. He turns around to look at the cave wall again, but he is not able to go back to his old way of thinking. He is caught between two worlds, having glimpsed a partial truth about reality.

Then Plato expands the story: With great trepidation, the man decides to step out of the darkness of the cave. As his eyes adjust to the sun, he sees the profound beauty of the real world: the colors, the shapes of the trees, and the village that lies in the distance. He now understands the difference between the small light of the campfire—the little "e"—and the profound light of the universe—the big "E" experience of Enlightenment. He realizes that he is witnessing a greater truth about the world. The metaphor is clear: We spend most of our lives experiencing only the shadows of reality. But if we can free ourselves from the assumptions and beliefs we hold in our mind—our cave of ignorance—we can become enlightened, first in small ways and hopefully in a life-transforming way. This, according to Plato, is wisdom, the highest form of awareness and self-realization we can reach.

Others have encountered similar experiences throughout history. Archimedes, the ancient Greek mathematician, exclaimed

"Eureka!"—"I have found it!"—when he discovered a profound scientific principle. I would consider this a small "e" experience: it provided great insight but did not fundamentally change his life.

For Buddha, his search for a way to end the suffering of others was a long and arduous path. When he finally reached a point of complete inner clarity—his big "E" experience—he gave up all selfish desire, declaring "I am awake."

DEFINING ENLIGHTENMENT, LARGE AND SMALL

We suggest that there is a spectrum of enlightenment ranging from little "e" to big "E" experiences. But what, *exactly,* is enlightenment? Perhaps the easiest way to define the little "e" experience is in the term itself: to shed light on our ignorance and bring ourselves out from the dark. The partial insights and epiphanies we have change our beliefs in small ways, often preparing us for the rarer big "E" experience where our entire worldview and values are radically transformed. Thus, for many philosophers and spiritual leaders, Enlightenment is perceived as the highest experience an individual can attain. Enlightenment also seems to be a universal phenomenon, with exemplars found in many cultures around the world.

Our research shows that when people have sudden spiritual or mystical experiences, they often describe a state of consciousness where everything feels deeply interconnected. These can be powerful, but for most people, they are brief. The big Enlightenment is typically associated with a permanent shift of perception, awareness, and knowledge. For some, the separation between God and one's self completely dissolves. For others, they feel a sense of absolute oneness with life, nature, or the universe. And for nearly everyone, the experience often feels *more* real than anything else in the world. "Truth" has been discovered, "God" has been touched, and insight has been gained. Are these little "e" or big "E" experiences? From a scientific perspective, we cannot say because it is the individual's *subjective* assessment—or the *opinion* of others—that dictate what "size" of enlightenment has taken place. In fact, when it comes to spiritual enlightenment, there will always be naysayers who will accuse the enlightened person of being delusional. Perhaps that explains why so many orthodox religions have persecuted those mystics who claimed to see a deeper or larger truth.

But the term still causes confusion. For example, Westerners often associate Enlightenment exclusively with Eastern religions, not realizing that there are powerful Jewish, Christian, and Muslim meditations that can profoundly deepen one's connection to God. Nonreligious people may also shy away from the term, not realizing that Enlightenment can be a secular experience, one that gave birth to the Western democracies of the world. And of course, there are many people who believe that enlightenment is nothing more than a fantasy or delusion. Because there is so much cultural and societal confusion concerning this nebulous term, I'd like to briefly describe some of the historical highlights.

EASTERN MYSTICAL ENLIGHTENMENT

Many Eastern philosophies define Enlightenment as the highest level of consciousness a person can attain. In Hinduism, consciousness itself is seen as the essence from which the universe emerged, and Enlightenment means that you have become one with this fundamental reality. In Taoism, you achieve Enlightenment by being in harmony with the principles of nature—with the "flow" of life. In such Eastern cultures, the language of Enlightenment is often put into poetry or phrased in paradoxical ways because it is so difficult to describe with words. For example, the Taoist sage, Laozi (Lao-Tzu) wrote, "Try to change something and you will ruin it; try to hold on to something and you will lose it."[1]

In Chinese and Tibetan Buddhism, Enlightenment is more personal, brought about through a process of continual self-reflection. In the Japanese Zen Buddhist tradition, students reach Enlightenment only when they realize the radical truth that everything is an illusion of the mind. Such an understanding is often accomplished by focusing on Zen koans (paradoxical stories or statements), which are used to interrupt the student's normal way of thinking. A person, for example, might be confronted by questions like these: "What is the sound of one hand clapping?" or "What is your original face?" Any logical answer could bring a blow from the master's stick!

Over the centuries, many of the Eastern Enlightenment philosophies became watered down and turned into folk religions filled with deity worship and tribal beliefs, and by the late nineteenth century, some of these religions began to fade away. In the twentieth century, many scholars began to introduce contemporary versions of the ancient Eastern sacred texts that appealed to a new generation of Westerners searching for spiritual Enlightenment. These seekers embraced a never-ending stream of gurus imported from Asia. Many people experimented with mind-altering drugs and unusual rituals borrowed from various spiritual or mystical traditions such as South American shamanic journeys, Sikh chanting, Sufi dancing, or Native American vision quests.

Some Westerners, after exploring these esoteric practices, took up research in medicine and psychotherapy. The result: spiritual disciplines like meditation and yoga were transformed into highly effective stress-reduction strategies that are now taught at universities and hospitals throughout America, a trend that *Time* magazine recently dubbed "The Mindfulness Revolution."[2] Stripped of their theologies, these practices are now taught in schools and are rapidly being integrated into corporate environments. In other words, even if we do not recognize them as such, Eastern perspectives of Enlightenment have become an integral part of mainstream America.

WESTERN MYSTICAL ENLIGHTENMENT

Whereas Enlightenment is deeply embedded in the philosophies of the East, the concept is rarely found in traditional Jewish, Christian, or Muslim sacred texts. This makes it particularly difficult for Westerners to fully grasp what the big "E" Enlightenment is all about. After all, there is no mention of a union with God in the Hebrew Bible, and the notion of becoming one with Jesus would probably be viewed as heresy. In these traditions, the "otherness" of God was emphasized. More emphasis was placed on faith and following the laws or commandments of the biblical texts.

Gnosticism, however, was an exception. Here the emphasis shifted from the *knowledge* of God to the *experience*—or mystical union—with the spiritual forces of the universe. The early Christians used the term "gnosis" to mean "knowledge by experience."[3] But the concept predates Christianity by many centuries. In fact, the notion of revelation—of being enlightened by the divine truth of the spiritual dimensions of the universe—was very popular between 300 BCE and 600 CE, a period that saw the rise of many prophets and the establishment of many religions throughout the Middle East.

It was also an era filled with great persecution by competing religious and political forces. Take, for example, the spread of Manichaeism, founded in Persia by a Babylonian named Mani, who came to be known as the "Apostle of Light" and supreme "Illuminator."[4] Like Muhammad (who was born three hundred years after Mani), he claimed to be the successor to a line of prophets that began with Adam and included Buddha, Zoroaster, and Jesus. His cosmology was similar to stories found in the Dead Sea scrolls, dating back to 300 BCE, in which the forces of light and dark constantly battled with each other. Thus, Enlightenment symbolized the release of the imprisoned light within each human being that allowed one to become free of evil and reunited with the Father of Greatness and the Mother of Light. Notice the similarity that these ancient stories have to Plato's allegory of the cave, where people are imprisoned in the dark and separated from the light of the heavenly world!

Manichaeism quickly spread through Europe and made its way into China and Tibet, demonstrating that there was a continual exchange of enlightenment theologies that deeply influenced early Jewish, Christian, and Muslim beliefs. Even St. Augustine, in his *Confessions*, stated that he first embraced and then rejected Manichaeism to form the Christian doctrines we are most familiar with today. This led to the persecution and eventual extinction of many of the European and Middle Eastern Enlightenment traditions that existed at that time.

In the Middle Ages, between the twelfth and sixteenth centuries, practices that encouraged the mystical union with God began to flourish in the esoteric traditions of Jewish Kabbalah and Islamic Sufism. In mystical Christianity, the concept of being in the *presence* of God was favored over the notion of becoming *one* with God. For example, an anonymous text called *The Cloud of Unknowing* was written in the mid-1300s as a guide to seekers who truly wanted to know God, and as the title states, you have to surrender your intellect if you want to experience the transcendent presence of God. In the 1970s, this early Christian contemplative practice, which was called Centering Prayer, was reintroduced to contemporary Catholic communities. However, the term "enlightenment" still is rarely found in Western theological discourse.

WESTERN RATIONAL ENLIGHTENMENT

In Europe, the idea of enlightenment took a very different path, one that eventually distanced itself from all things spiritual, magical, or supernatural. The term first appeared in the late fourteenth century, and was a reference to a person who was illuminated, well informed, or educated. But as the Renaissance spread across Europe, a bevy of esoteric practices filtered in from various Eastern, Middle Eastern, and Russian traditions. The markets were filled with purveyors of alchemy, astrology, numerology, mediumship, and Tarot divination, and a large assortment of mystical teachers claiming to unveil the mysteries of the universe. Each group promised its own brand of "Enlightenment."

The Catholic Church reacted to these "heretical" movements with persecution and over one hundred thousand people purportedly died as a result. What brought this slaughter of humanity to an end? The emergence of a new philosophical movement called the Age of Enlightenment, also known as the Age of Reason. It was an ideological war against the popes and against the politics of the time, and it gave birth to some of the greatest intellectuals, artists, and religious reformers in Western history: Bacon, Spinoza, Locke, Hume, Descartes, Newton, Voltaire, Rousseau; and in America, Franklin and Jefferson.

Enlightenment was redefined. Immanuel Kant defined it as

the emancipation of human consciousness from a state of ignorance[5] and others defined it as freedom *from* religion. History
professor Dorinda Outram summarized the competing perspectives succinctly:

> Enlightenment was a desire for human affairs to be guided
> by rationality rather than by faith, superstition, or revela-
> tion; a belief . . . validated by science rather than by reli-
> gion or tradition.[6]

The big "E" Enlightenment described in Eastern philosophy
was reduced to the "aha" experiences associated with rational
thinking, contemplative self-reflection, and scientific insight.

AMERICAN SPIRITUAL ENLIGHTENMENT

The attempt to eliminate supernatural beliefs did not succeed.
Instead it gave birth to romanticism, from which a new form of
enlightenment was created: transcendentalism. One became enlightened by immersing one's self in the joys of the *human* spirit,
epitomized by a love of nature, sensuality, and the arts.

Eastern concepts of Enlightenment became popular in Europe,
and rather than being destroyed, mysticism went underground,
eventually to reemerge in full force in nineteenth-century America. Transcendentalism was quickly absorbed by divinity schools
like Harvard and could be found in Unitarianism, Christian Science, and a variety of charismatic and New Thought movements

that encouraged congregants to become *one* with God, universal consciousness, and the Holy Spirit. Such experiences can have the life-transforming consequences of either a big "E" or little "e" experience.

By the early 1900s, many Eastern views of Enlightenment filtered into mainstream Christianity and permanently changed the landscape of American religion and spirituality. Suddenly both little "e" and big "E" experiences were available for anyone and everyone who sought to become enlightened through God. To put it another way, we can all become "awakened" if we delve deeply into our spiritual beliefs.

PSYCHOLOGICAL ENLIGHTENMENT

In the late 1800s, Enlightenment also became a popular topic among North American doctors and psychiatrists. It began with William James, one of the founders of American psychology, who collected stories of spiritual awakening. In his book *The Varieties of Religious Experience,* he quotes Tolstoy's memoir describing his struggle with depression and his sudden Enlightenment:

> I felt that something had broken within me on which my life had always rested, that I had nothing left to hold on to, and that morally my life had stopped. An invincible force impelled me to get rid of my existence, in one way or another. . . .
>
> I remember one day in early spring, I was alone in the

forest, lending my ear to its mysterious noises. I listened, and my thought went back to what for these three years it always was busy with—the quest of God. But the idea of him, I said, how did I ever come by the idea?

And again there arose in me, with this thought, glad aspirations towards life. Everything in me awoke and received a meaning. . . . "Why do I look farther?" a voice within me asked. He is there: he, without whom one cannot live. To acknowledge God and to live are one and the same thing. God is what life is. Well, then! Live, seek God, and there will be no life without him. . . .

After this, things cleared up within me and about me better than ever, and the light has never wholly died away. I was saved from suicide. Just how or when the change took place I cannot tell. But as insensibly and gradually as the force of life had been annulled within me, and I had reached my moral death-bed, just as gradually and imperceptibly did the energy of life come back. And what was strange was that this energy that came back was nothing new. It was my ancient juvenile force of faith, the belief that the sole purpose of my life was to be better. I gave up the life of the conventional world, recognizing it to be no life, but a parody on life, which its superfluities simply keep us from comprehending.[7]

Tolstoy had what William James called a "conversion" experience: a gradual epiphany that led him to first question and then reject institutionalized religion, which he believed corrupted the message of Jesus.[8] Like Descartes, Tolstoy questioned everything: logic, reason, knowledge, and especially the sacred texts of Christianity. And yet God, *the infinite*, is what he eventually found, a presence that if correctly understood would "let us love one another in unity."[9] After his Enlightenment, Tolstoy would devote the next years of his life to rewriting the doctrines of Christianity.

James highlighted this story because it contained a common thread: when people experience Enlightenment, old religious doctrines often appear to be false. For James, this was more than just a religious epiphany; it was a *psychological* transformation of the human personality.

Around the same time James was preparing his famous lectures, a distinguished Canadian psychiatrist named Richard Bucke published a description of his own Enlightenment experience, which gave him a new vision of reality:

All at once, without warning of any kind, I found myself wrapped around as it were by a flame-colored cloud. For an instant I thought of fire, some sudden conflagration in the great city; the next, I knew that the light was within myself. Directly afterwards came a sense of exultation, of immense joyousness immediately followed by an intellectual illumination quite impossible to describe. Into my brain streamed one momentary lightning-flash of the Splendor which has ever since lightened my life; upon my heart fell one drop of Bliss, leaving thenceforward an

> aftertaste of heaven. . . . The illumination itself continued
> not more than a few moments, but its effects proved inef-
> faceable; it was impossible for me ever to . . . doubt the
> truth of what was then presented to my mind.[10]

Due to this experience, Bucke devoted the rest of his life to helping mentally ill patients. Whereas Tolstoy was led to reject religion, Bucke—a nonreligious man—found himself driven to explore the spiritual and mystical traditions of the world, culminating in the writing of one of the most famous books of the twentieth century, *Cosmic Consciousness*, published in 1901. Together with *The Varieties of Religious Experience*, these two books opened the door to many formal academic studies into the nature of spiritual and transcendent experiences.

James summarized the value of Enlightenment in this famous quote: "There are two lives, the natural and the spiritual, and we must lose the one before we can participate in the other." James was also the first person to insist that these altered states of mind "are neurally conditioned," an intuition that our brain-scan studies confirm. Bucke, like James, attempted to identify the common elements of the "cosmic consciousness" process, which included:

- a subjective experience of "inner light"
- a deepening of moral or spiritual values
- an increased sense of intellectual illumination
- a feeling of immortality or eternity
- a loss of the fear of death

- a loss of the sense of sin or guilt
- an instantaneousness awakening
- a lasting transformation of personality

The last concept specifically refers to the big "E" experience of Enlightenment: You are no longer the person you used to be. You are transformed. Your character has changed. Your values have changed. The way you see yourself in relationship to everything has changed, and this often leads to a new direction and purpose in life.

Other psychologists have also built models to describe the qualities of enlightenment. For example, Abraham Maslow proposed a hierarchy of human needs, culminating with a level of enlightened consciousness he called "self-realization." He, like James and Bucke, argued that Enlightenment is a natural, developmental, and biological state of mind. Other researchers have expanded on James's and Maslow's definitions to include feelings of oneness with nature, loss of space or time, dissolution of self-identity, and a sacred sense of life.

ENLIGHTENMENT AND YOUR BRAIN

Big "E" or little "e" experiences both can have a profound effect on the brain. With brief moments of insight (the "aha" experience of little "e"), we can record temporary changes in many neurological areas. However, the research we and others have accumulated suggests that many small "e" experiences can lead to long-term changes in the brain, changes that actually affect the way we see reality. No one can promise you that you'll experience the big "E," but perhaps that's not the point. What is important is to realize that we can be on the path *toward* Enlightenment, with the evidence that every step along the way improves the functioning of your brain.

For example, we have gathered evidence showing that the quest for enlightenment—large or small—causes long-term changes that affect the emotional and cognitive centers of the brain. That is why a person feels and thinks differently when they experience a dramatic shift in consciousness.

Normally, we think that the brain changes slowly. It takes time for the brain to learn new skills and absorb all of the experiences we have in meaningful ways. But the Enlightenment process suggests that your brain can change in an instant. How is that possible? We will explore this in later chapters, but part of the answer is that it is already in you to begin with. It is just a matter of finding it.

TAKING PICTURES OF ENLIGHTENMENT

The current chapter in humanity's search for Enlightenment now takes place in neuroscientific laboratories throughout the world, in huge magnetic donuts that can map the neurological activity of mystical experiences as they occur in real time in the meditator's brain. Now we can see which areas of the brain are affected, and measure the benefits or drawbacks that come when such experiences occur.

We know that gentle contemplative practices like mindfulness meditation predict an improvement in one's mood, empathy, and self-awareness. But Enlightenment is something else, marked by a sudden and intense shift in consciousness. Even if you deliberately seek it, you may or may not find it. For one person, it happens overnight, but for another, years or decades may pass.

There's a wonderful story about a young man who goes to a Buddhist monastery eager to become enlightened as quickly as possible. The man asks the Buddhist master, "How long will it take me to become enlightened?" The master answers, "About ten years." The young man says, "Ten years! Why ten years?" The master replies, "Ah, I see that you are right! In your case, twenty years!"

"Why do you now say twenty years?" the frustrated man retorts, but the master then says, "Oh, I'm sorry. I was mistaken. For you, thirty years."

In other words, the more you *try* to become enlightened, the more it may elude you, which is exactly what happened to me. I had to give up my "outer" quest for truth and trust some inner or higher process to take over, which led me to my experience of Infinite Doubt. It's like trying to see in the dark: the harder you stare, the less you see. But if you surrender yourself to the unknown—if you stop trying to use your eyes and open up your other senses—you may be able to see the situation in a new "light," free from your memories and old beliefs.

For some, Enlightenment may occur when they reach the end of their rope, or when they are in a life-threatening situation. It forces them to completely reevaluate their life. For others, smaller enlightenment experiences may come when they experience a major life change, like becoming pregnant or seeing your newborn infant for the first time. But the most fascinating ones are those that seem to come out of nowhere. Here's an example reported on our online survey by a fifty-one-year-old nonreligious physician. All he was doing was driving a moving van to help a friend:

> As the truck was crossing a long bridge, I suddenly noticed some kind of shift in my awareness. The boundaries between the bridge, my truck, and myself began to blur, and the thought that came to mind was this: "It's all the same, no different than you."
>
> As I continued down the road, every single object that I viewed became "the Same." Everything had a Sameness. (I use capital letters because the experience had a profound and sacred quality.)
>
> I was filled with an aura of delight, and the feeling kept rising and rising. I observed my own mind and I *experienced* my thoughts as though they were dominoes falling, one after the other, in a long line. I watched as all my "problems" vanished, not because any of them were solved, but because the questions themselves disappeared. I thought to myself, "There was never any problem in the first place." I was just lost, and now I'm found. I am All of It, this sacred Sameness.

A few minutes later, the Understanding began to fade. And yet this experience permanently changed my life. It felt more real than what I usually think of as real. Truth was unveiled, and I realized how futile it was to seek happiness in the old reality that I used to live.

This is an outstanding example of a spontaneous experience filled with powerful emotions and a sense of unity and permanence. The experience came out of nowhere and caused this person to have a dramatic new perspective on the world. Enlightenment, as this example shows, often brings about a new sense of clarity that can change the entire belief system of the individual.

With well over a hundred years of research, there emerges a consistent pattern, one that allows us to view enlightenment—little or big—as a process of personal growth, and each of us will interpret these "awakenings" in different ways. Large or small, spontaneous or gradual, each new insight allows our consciousness to grow.

THREE

What Enlightenment Feels Like

In mystic states we both become one with the
Absolute and we become aware of our oneness.
This is the everlasting and triumphant mystical
tradition, hardly altered by differences of clime or
creed. In Hinduism, in Neoplatonism, in Sufism, in
Christian mysticism, in Whitmanism, we find the
same recurring note.

William James, *The Varieties of Religious Experience*[1]

While there are many well-known stories describing experiences of Enlightenment in the world's spiritual literature, I wanted to find out how many people in today's world had experienced something similar. So I created the online survey mentioned earlier to explore the many facets of these remarkable events.

I encouraged people to write about their most powerful spiritual or transformational experiences, and by using this strategy, I gathered nearly two thousand remarkable stories. This allowed me to compare both the commonalities and the uniqueness of their little "e" and big "E" experiences and to measure the psychological and spiritual effects of enlightenment. What, for example, do people feel and think when they experience a profound shift in awareness? Do age, gender, or finances influence the likelihood of having such an experience, and is there a specific "button" we can push to evoke them? From the data I collected, I could begin to draw connections to how consciousness and the brain can be affected by the experiences that lead to Enlightenment.

THE SPIRITUAL EXPERIENCE SURVEY

When investigating the nature of Enlightenment, one of the biggest problems is language. In fact, I found that most people in our survey struggled to find the right words, often using convoluted metaphors, quotes, and capital letters ("Life is a branching tree," "FORCE" instead of "force," etc.) to capture an experience that words failed to explain. To address that problem, I encouraged our survey participants to freely describe, in their own words and in their own way, the types of spiritual experiences they had. But I also realized that it was important to ask specific, pointed questions that could help people distinguish between the big "E" and little "e" experiences.

I wanted to get at the very essence of the transformation that was produced in that person's life. For example, did it remove suffering? Did it change his entire belief system? Did it alter the trajectory of her relationships, religion, or career?

I designed a website where we could ask the kinds of questions that other researchers had not done: "How did you feel before and after the event?" "What emotions did you have?" "What behaviors were changed as a result of your encounter with this rare state of consciousness?" "Was it relaxing or stimulating, peaceful or disturbing?" And most important, I asked them how "real" the experience felt to them.

I posted the survey online, and wherever I went I encouraged people who had enlightenment-like experiences to provide us with detailed information about their lives. People from many spiritual traditions were asked to participate—students, teachers, corporate executives, etc.—and we received responses from around the world.

Participants also filled out a variety of standardized questionnaires concerning their background, religious beliefs, spiritual feelings, and ideas about death and how much they were still searching for answers. I collected information about their childhood and family beliefs to see if those beliefs were altered after they had their epiphanies. I also inquired if the experience happened under the influence of a hallucinogenic drug or during a near-death experience, a religious conversion, or intense meditation or prayer.

In designing the survey, I also collected information that I felt could be directly correlated to specific brain processes that have been identified through our brain-scan studies on spiritual practices. For example, if someone described incredible joy, I could assume that there was intense stimulation in the positive emotion areas of the brain. Or if a person felt an intense sense of unity with

God, her parietal lobe should be affected since that part of the brain regulates our sense of oneness and connectedness with others. If a person was suddenly filled with feelings of love and compassion, it's highly likely that we would see increased activity in the areas of the brain associated with insight, social awareness, and our positive feelings toward others.

THE ELEMENTS OF ENLIGHTENMENT

I specifically wanted to know if religious background and current beliefs affected the types of experience a person had. After all, if everyone who answered the survey was a churchgoing Catholic, it wouldn't tell us if Buddhists or atheists had different experiences. So I was pleased to discover that we had expansive diversity across many demographics. Some were rich and some were poor, and our survey captured people from many ethnicities. Most of the respondents were American, but 15 percent lived in other parts of the world. We had equal numbers of women and men, with ages ranging from eighteen to eighty-two, and although many were mainstream Christians (Catholics and Protestants), we had many Jewish, Muslim, Hindu, and Buddhist participants. There was also a large atheist contingent—about 25 percent—which is consistent with other surveys of religious preference. Many of our respondents described themselves as agnostics, spiritualists, or those who blend different religious traditions into formal and informal practices.

So what did we find? My analysis led me to the groundbreaking conclusion that there are five basic elements that lead to an Enlightenment experience, and they are generally the same for everyone:

1. A sense of unity or connectedness.
2. An incredible intensity of experience.
3. A sense of clarity and new understanding in a fundamental way.
4. A sense of surrender or loss of voluntary control.
5. A sense that something—one's beliefs, one's life, one's purpose—has suddenly and permanently changed.

However, the interpretation of these elements varies enormously from person to person. For example, you might feel unity with nature, universal consciousness, or God. Here is a description from a sixty-five-year-old American Jewish woman:

It felt like an energetic merging and being at One with the most powerful Creative Force/Being in and beyond all universes. In that moment, I was simultaneously the same individual consciousness of myself, but I was also a part of "God" (for lack of a better term, really). Infused with the power of all Creation/Creativity, I was buoyed with a joy so immense it infused my Beingness with an affinity for everything.

Now notice the similarity in the sense of unity with this description provided by a forty-three-year-old woman from India:

> Once when I was practicing a Pranic Healing [sending energy to parts of the body through one's hands] tears were streaming down my cheeks. Then I experienced a feeling of "ONENESS" with all beings. My body felt very light and there was no separation between me and the external reality. There was no sense of "self." It was unique.

Both of these examples describe feelings of unity—a sense of Divine connectedness with God in one case, and a connection to all living beings and reality itself in the other. But the second story exemplifies another quality that is often described in Eastern scriptures: a loss of the sense of self. The person literally feels as if her own self is dissolving. There is no "I"—just the totality of a singular awareness or experience.

Think about that for a moment. How would you feel if you had the experience that your own sense of self was disappearing? It sounds like a very scary experience, because the brain uses self-identity to make distinctions between ourselves, other people, and the world. If there was no "I" or "you" or "we" or "it," everything would blend together. There would be no separation, and the human brain usually finds that type of experience very disorienting. It also becomes very difficult to use descriptive language, because words help the brain to create separations between our bodies and the outside world.

I had the same feeling during my experience of Infinite Doubt. After all, even my own self was doubted. I felt like I lost all connection to my own identity as I became profoundly connected to everything else. Interestingly, it's the same kind of feeling reported by the Buddhists and Franciscan nuns who participated in our brain-scan studies. At the moment they experienced a sense of oneness or a loss of self, we observed a sudden drop of activity in the parietal lobe, the area that creates an arbitrary distinction between "self" and "other." The parietal lobe helps us to orient ourselves to external objects, so a decrease of activity coincides with a decreased sense of self. Boundaries are blurred, giving us a sense of connectedness with everything—God, nature, the universe, etc. It doesn't appear to matter what a person's religious or cultural background might be. We all have the neurological ability to feel this powerful sense of connectedness and unity that is usually associated with big "E" Enlightenment.

The second element of the Enlightenment experience, intensity, is often related to the sense of oneness but might be related to other components of these experiences too. A person might suddenly feel profound joy or love, or might see an intense bright light or hear a sound that seems extraordinarily beautiful. Take this example from a forty-three-year-old man:

> I, as an un-namable but individual being, was traveling
> down an infinite roller coaster like waves of pure white
> ecstatic light. The ecstasy was overwhelming and rose and
> fell in intensity with the waves of light. The light path

seemed infinitely long in both directions. The sense of the
being and the light was INFINITELY MORE REAL than
anything I had ever experienced.

You can see how he feels the need to describe his experience using
the emotionally powerful words: "ecstatic," "overwhelming," "infinite," "pure." Such words help the individual impart the dramatic
intensity of the experience, and most of the respondents used
such language to describe their experiences. Profound intensity
seems to propel the experience from a little "e" to a big "E"
transformation by making it more real.

A third major element that occurs in the Enlightenment experience is a sense of great clarity. Many people said that the experience changed their life because everything now seemed to make
more sense in their lives. They more clearly understood their purpose in life, the value of relationships, and the goals they wanted
to set for their future. This is what occurred to a thirty-seven-
year-old female scientist during a deep meditation session:

Everything in life seemed to click. I had this clarity and it
was as if I was looking at life from the inside out. I work in
science, and I grew up in a conservative religion, but I have
always rejected and tried to avoid the idea of "blind faith."
Despite my trepidation, this experience seemed to satisfy
my proof-oriented mentality with the concept of intuition.
It was almost as if my intuition from somewhere "deeper"

had offered some sort of direct experience that offered
up proof.

This new sense of clarity in her life transformed the way she
thought, felt, and acted: "I even feel it has made me a better scientist and thinker, because I have less fear of the truth."

The fourth crucial element of Enlightenment is the sense of
surrender to the experience as it unfolds. In other words, *you* are
not directing the experience or making it happen; it is directing
you. You are going along for the ride. Think about the term
"Enlightenment" again—where does the light come from? Like
sunlight, it comes from a source different from you. Take a look at
this powerful description of surrender described by a Catholic
woman:

> I was in anguish. I was lost and I had no sense of God's
> direction. Everything was dark. I cried out, but nothing
> came to me. Suddenly I had the experience of God asking
> me if I would do anything He asked. This was not an audi-
> ble voice, rather a knowing inside. I said "yes," but was met
> with silence. Another day passed and then He asked if I
> would be willing to give up everything for Him, even my
> religious faith and salvation. That took me aback. I couldn't
> believe He would ask that of me. So I waited and tried to
> discern if it was God who was asking me that or some
> other spirit. I prayed for another couple of days, and the

anguish increased. Finally, I decided I could not withhold my assent. I *surrendered* everything, including my faith and my salvation, and only for one reason. I loved God so much that I would truly give up everything to be connected with Him. I said "yes" and in an instant, God returned everything to me, transformed. He liberated me. From that day forward, a new relationship exists between God and me. It is ever present, no distance, no separation. It is! How has it changed? I am not attached to doctrine, dogmas, or rituals. I see God's action all around me.

The moment she surrendered herself up to something that challenged her core beliefs she received something greater, an experience that reflected many of the elements of Enlightenment. In this case, her entire religious orientation permanently changed as a new relationship with God was born. Old beliefs fell away, and her entire worldview was altered: she saw God everywhere.

The sense of permanent change is perhaps the most important element because it more definitively distinguishes the little "e" experiences from the big "E" encounters that transform one's personality. For example, a person can have an intense emotional experience from an orgasm, creating a momentary sense of oneness or unity, but it doesn't change the direction of that person's life. You can work in a science lab and have a profound insight that gives you intense *clarity* about your work, but you probably wouldn't feel that you were enlightened for more than a minute or two.

Not everyone in our survey made reference to all five elements of Enlightenment, and many people did not clearly state how much the experience changed their lives. So it is difficult to say whether each of these events were small "e" or big "E" experiences. In fact, even the most profoundly life-changing events often left the respondent feeling rather humble. Remember that even Buddha did not claim to be Enlightened, only that he was now awake. But he did experience oneness, clarity, intensity, and surrender. And yes, his life was permanently changed.

THE UNIQUENESS OF ENLIGHTENMENT

When I did a content analysis of the participants' descriptions, I was astonished to discover the wide range of variation even among people of similar backgrounds. For example, considering that most of our respondents were religious, only 18 percent of all the narratives mentioned God, and less than 4 percent mentioned Jesus, even though half the religious respondents were Christian. Barely 10 percent mentioned love, and less than 5 percent referred to consciousness or truth—concepts that I would have expected to be referred to far more often when describing transformational experiences. In essence, everyone described their experiences in unique ways.

However, I did find gender patterns. Men's experiences were

focused more on the world, the universe, and consciousness, while women focused more on God, love, relationships, and children. The following two examples exemplify this:

> When my third child was born, I experienced prayer in a way I've also not experienced again. I felt such overwhelming joy that for about thirty-six hours—interrupted only by sleep—I couldn't stop praising God in my mind. I felt as if I were being carried on a thrilling current I could've stepped out of. Only I didn't want to step out. This feeling of ecstasy in praise finally wore off slowly, as if I were descending gradually back to a normal state. I think being with God in heaven may be like what I experienced in this ecstasy.

For this woman, her child's birth triggered an intense spiritual experience, but it clearly doesn't have the hallmarks of Enlightenment since it appeared to wear off and not result in a permanent change in her overall beliefs and behaviors. However, we believe that these small "e" experiences can pave the way to future life-changing events.

Now compare this story from a seventy-year-old male college professor:

> I am not so much interested in joining or being a part of a spiritual group as I am in conducting participant observation of their rituals in the context of their shaman-

istic beliefs. I have been profoundly impacted by experiences of altered states of consciousness to the point where I would say that my religion is more of a metaphysical philosophy.

These two representative examples show that men are typically more globally or "large scale" oriented while women are more "community and family" oriented. Or to put it another way, men focus on abstract ideals and women focus on interpersonal relationships. Of course, this is not always true, and in our survey, there were many women who referred to the "big concepts" and many men who reflected on their family. But by and large, gender differences tended to influence the content and direction of Enlightenment experiences.

Every person has a different neurochemical and hormonal "profile," and every human brain—male and female—functions in slightly different ways. Even though men and women may interpret Enlightenment differently, the benefits and end results are largely the same. Everyone feels that the experience transforms their life, and we still find the other essential elements of unity, clarity, intensity, and surrender.

Another common denominator was the use of words such as "force," "energy," and "power." They all reflect qualities associated with God and spirituality, and yet the lack of typical language suggests that Enlightenment transcends scriptural concepts of a human-like deity. This distinction reflects another significant finding: intense personal spiritual awakenings appear to weaken the doctrinal religious beliefs our participants were raised with. The Catholic woman's description above reveals how she moved away from tradition and dogma to a very different sense of spirituality.

Enlightenment can also cause a person to entirely abandon religion.

For most people, Enlightenment experiences deepened their interest in spiritual pursuits. In fact, there was a 10 percent decline in participants' *religious* interests and beliefs, but an 89 percent increase in the participants' *spiritual* activities. Many respondents felt that their experiences were not adequately addressed by the religions in which they were raised, and so they turned away from traditional doctrines to engage in more individualized pursuits. Still, more than 50 percent of those who strongly identified themselves with a specific religion felt that their faith was deepened, and for nearly everyone, they felt that their lives had new meaning and purpose.

THE REALNESS OF ENLIGHTENMENT

Normally, we take reality for granted. Have you ever looked at your house or your car and questioned whether it was an illusion? Probably not! In fact, why do you *assume* that the book you're holding is real? For the most part, your brain doesn't care if something is real, it only cares if the map—the vision, the sound, or even the fantasy—is *useful.* Even when we dream, the brain does not question the reality of the experience at the time; we only do that when we wake up, because dreams are not generally useful for dealing with daily tasks.

When I was young, I used to have a recurring dream in which I was being chased by a dinosaur. I knew that dinosaurs no longer existed in the world, but in my dream state I would run away so fast that when I woke up I was literally out of breath. My brain treated the dinosaur as if it was real. But after a minute or two, my new wakeful state of consciousness would say to itself, "Oh, that was *just* a dream."

Enlightenment is different. None of our survey participants said, "It was a neat experience, but it was nothing more than a fantasy." Just as our everyday reality feels more real than a dream, Enlightenment feels more real than everyday reality! A sixty-year-old psychiatrist put it this way:

> My experience felt utterly real, more real than usual. I felt like I tapped into something very ancient, powerful, and connected; something that I rarely experience in day-to-day life. And it's only in these brief periods of transcendence do I really feel alive. They feel more real than life itself.

My survey included a specific question asking how "real" the experience seemed as opposed to being a fantasy or hallucination, and the response was nearly unanimous that the encounter was extraordinarily real. Interestingly, neuroscientists like myself have never been able to crack this issue: Why do we think something—anything—is real? And what part or parts of the brain help us to make this determination? Our brain-scan research suggests one possibility. In many of our studies, we have noted that spiritual practices and intense experiences have a powerful effect on the thalamus, a central brain structure that processes sensory information and helps different parts of the brain communicate with one another. The significant changes we see in the thalamus during profound spiritual states may explain why Enlightenment feels so extraordinarily real.

There is another piece to this reality puzzle that is unique to an Enlightenment experience. Unlike a dream, which suddenly seems unreal after you wake up, an Enlightenment experience still seems *more* real than everyday reality, even when you look back on it years later. It's like saying that the dream you had last year continues to feel more real than the rest of your life. How strange and incredible, and yet over 90 percent of our survey respondents reported that their experience continued to feel as real or more real than the reality they normally experience every day.

This speaks to the transformational effect of Enlightenment since it appears to stay with you for the rest of your life. And for most people who have this experience, there is great comfort knowing that there is a "greater" or "truer" reality than the one perceived by the brain. For most of our respondents, that deeply felt sense lessened their worries and fears. That's one of the things

Enlightenment does: it makes our suffering, or the reason for our suffering, feel less real.

THE PERMANENCE OF ENLIGHTENMENT

As we stated earlier, the hallmark of the big "E" experience is permanence. The event changes major aspects of our life: our values and belief systems, even our habits and behavior.

Take this wonderful example of an experience that permanently transformed this fifty-three-year-old gentleman's life:

> During an "altar call" the church members gathered around as the ministers laid hands on me and prayed for my deliverance. I felt a spiritual presence and heard a voice speaking in my mind. The moment I made the mental decision to connect with this entity, I felt a transformation take place in the area of my heart, an area that felt hard and unrelenting. In the past, I could commit acts that were both illegal and immoral, but I never felt remorse or shed a tear. On the night of my deliverance, the "hardness" crumbled and I felt what I can describe only as "warm oil" flowing over my inner being. From that moment, my daily thoughts began to change. I completely stopped smoking, even though I had tried to quit numerous times in the past. I gave up all illegal activities and I lost interest in all recreational drugs, even though it had been a way of life

for me. Months later, I decided to become a minister in the faith.

He had a change that overcame deep-rooted personality problems that few therapies could address. In a single day, he was completely changed and his addictions instantaneously melted away. Our survey is replete with descriptions of similar amazing experiences that transformed a person's behavior.

As you can see in the chart on the next page, for 80 percent of those who experienced moments of Enlightenment, their meaning of and purpose in life were greatly enhanced. This often took the form of better interpersonal relationships with friends and family and a renewed enthusiasm for their career or doing other meaningful things with their life. Fear of death dramatically declined and 56 percent of the respondents felt that their health had also improved. They psychologically and physically felt better. Nearly 90 percent felt much better about their spiritual pursuits.

Enlightenment can improve virtually every corner of your life: your relationships, your health, and your desire to find more meaning and purpose. Once you've had a powerfully transforming experience, even the memory of the event is enough to reinforce the changes it evoked. Here is how this young man describes it:

> Whenever I reflect on these profound moments of insight, I feel like I tap into something very ancient, powerful, and connected; something I forget in day-to-day life. And it's

only in these rare and brief periods of transcendence that I really feel alive.

When people relive the feeling of Enlightenment, it reinforces the elements that made them change their life.

HOW HAS THIS EXPERIENCE CHANGED THE FOLLOWING					
	Much Better	Somewhat Better	No Change	Somewhat Worse	Much Worse
Family Relationships	32.8%	27.4%	33.5%	4.9%	1.5%
Fear of Death	55.3%	20.4%	23.1%	0.8%	0.4%
Health	27.9%	28.0%	41.7%	1.7%	0.7%
Purpose in Life	55.5%	25.3%	16.8%	1.9%	0.4%
Religiousness	27.3%	25.2%	37.0%	4.7%	5.8%
Spirituality	71.2%	18.0%	9.2%	0.8%	0.8%

FOUR

Enlightenment Without God

Western concepts of enlightenment usually refer to the little "e" experiences associated with the intuitive "aha" moments that give us insights into issues we are wrestling with, such as solving a scientific problem or figuring out how to resolve a complex interpersonal issue. They are important because they can prime the brain toward having the big "E" experiences that change one's life.

However, Enlightenment is usually seen as a form of *spiritual* development, and for people who do not relate to God, religion, or theological philosophy, the word itself can be a turnoff. This may cause a person to not realize that he or she may have actually experienced some or all of the five common elements of Enlightenment we identified in the previous chapter: oneness, clarity, intensity, surrender, and permanent change in some core aspect of one's life. Since our main purpose in writing this book is to show the neurological evidence that personal transformation is available to everyone, we want to address Enlightenment as seen through the eyes of a disbeliever.

The past decade has seen a dramatic rise in atheism, and religious affiliation is at its lowest point in American history.[1] In fact, over forty-six million Americans publicly declare themselves nonreligious.[2] That's 20 percent of the adult population, with nearly a 60 percent drop-out rate for those who are younger than thirty. These millennials are three times more likely to abandon the religion of their parents, and they are convinced that their spiritual needs can't be met by traditional faiths.[3] One young woman in our study captured this new generational attitude:

> Both my parents had died in my early twenties, and I had dabbled in drugs and promiscuous sex. I also was kind of adrift and was unsure what direction my life would take. At the time, I was agnostic when I experienced a gradual religious awakening. I kept having a strong feeling of being protected, that someone or something—God perhaps— was watching over me. Because of this, I tried to reenter the church of my childhood, but I realized it was definitely not for me. The dogma was oppressive. Long story short, I came to realize that I do not seem to fit into any kind of organized religion. It's the dogma thing—the "We're okay, you're not" mentality that bothers me.

Yet the quest for spiritual experience appears to be increasing. As a national study of college students showed, 80 percent say that spirituality is essential to their lives.[4] It's not scriptural doctrine that intrigues them, but the experience of feeling more fully alive and being connected to others and to the universe as a whole. Whether they know it or not, they too are seeking Enlightenment.

The data from our survey demonstrate that believers and nonbelievers alike can experience Enlightenment. In this chapter, we'll take a closer look at how the experience works outside a religious framework, leading to greater acceptance and tolerance between religious and nonreligious groups.

A GODLESS ENLIGHTENMENT?

Our survey reveals a fascinating perspective on the enlightenment experiences of nonreligious people. Many of our respondents labeled themselves atheists and yet they still described their experiences using spiritual terminology. As one person wrote:

> I am not at all religious, but I have experienced a variety of spiritual phenomena. I once felt a connection with a divine conduit which I experienced as a plurality of one mind, what people refer to as "God." The experience instantaneously answered all of the questions I had with regard to the nature of existence and religions in general.

Overall, nonreligious people define spirituality as a deep sense of connectedness and purpose that gives life meaning and fulfillment. For many, Enlightenment came while they specifically pondered the nature of religious belief:

I had a true spiritual awakening when I realized that religion and spirituality are completely unnecessary to my life and that I need not adopt any belief system other than an "existential" attitude. Nothing else makes sense to me and I believe that my peace of mind depends on being able to make sense out of the world I live in. I need no outside validation nor do I need a set of rules to adhere to in order to live a satisfying and good moral life.

Others came to the conclusion that religion was not relevant to understand the universe. Take, for example, this thirty-eight-year-old man who considered himself agnostic. He had a spontaneous experience that caused him to abandon all religious pursuits:

I was walking down the street when everything came "alive." The wind felt more real than before. The mountains looked stupendous. Everything had more depth, more feeling, and I felt a type of inner joy—or maybe it was peace—I had never experienced before. Now whenever I walk in nature, everything seems more beautiful and I feel a sense of fullness and contentment I have never known. In that moment I "knew" that god did not exist and that I had no reason to embrace religious values. I felt—and still feel—whole just being in the present moment.

Interestingly, the *quality* of the experience for an atheist is often indistinguishable from the most devout Christian, Muslim, or Jew. Atheists have *intense* experiences in which they feel powerful emotions and a profound sense of *clarity* regarding their understanding of life. They too describe a strong sense of *unity* with the rest of humanity or the universe, and yet the experience convinces them that God does *not* exist, as this survey respondent stated:

> Culturally, I remain Mormon to this day. Medical training in molecular biology caused me to believe that evolution was much more probable than creation by an intelligent designer, so I reexamined my past spiritual experiences in detail. I concluded that I had put too much faith in my own feelings, and that a Heavenly Father probably did not exist. I decided to pray about this insight, and I experienced that same intense warmth in my bosom that I received from earlier spiritual epiphanies. In effect, God "testified" to me that the Heavenly Father does not exist. This experience transformed my life.

With or without God, the Enlightenment experience can change a person's entire system of belief. For some, religion is completely abandoned for science, but for others, like Einstein, an underlying spirituality remained, embedded in a scientific worldview:

> The most beautiful and deepest experience a man can have is the sense of the mysterious. It is the underlying principle of religion as well as all serious endeavors in art

and science. He who never had this experience seems to me, if not dead, then at least blind.[5]

Einstein was deeply influenced by Western enlightenment philosophers like Baruch Spinoza and Immanuel Kant, and in his autobiography, Einstein described several childhood experiences of profound "wonderment" that changed and shaped the future direction of his life. The first occurred when he was about five years old, when his father showed him a compass:

> I can still remember . . . that this experience made a deep and lasting impression upon me. Something deeply hidden had to be behind things. . . . At the age of twelve I experienced a second wonder of a totally different nature: in a little book dealing with Euclidean plane geometry . . . [its] lucidity and certainty made an indescribable impression upon me.[6]

This suggests that Enlightenment experiences can occur quite early in a child's life, and these moments can shape a person's future career.

The atheists, like the rest of the survey participants, generally reported that Enlightenment left them with a greater sense of meaning and purpose in life, better interpersonal relationships, and a more optimistic view of life in general. The results suggest that Enlightenment has the following universal effects:

- They feel more open-minded.
- They don't ruminate on past mistakes.
- They are not as worried or nervous about facing and solving problems.
- They find themselves to be happier, more peaceful, and more contented with their life.

THE SECULAR LANGUAGE OF ENLIGHTENMENT

In general, atheists reflect deeply on religious and spiritual themes, but their Enlightenment experiences often cause them to feel more certain that religious beliefs are false. Of course, one could argue that Enlightenment causes everyone to challenge old beliefs and to revise or replace them with more meaningful concepts. These concepts can take the form of religious or nonreligious doctrines and often include concepts of love, compassion, or freedom. Thus, when you look below the surface, nonbelievers hold values that are often reflected in religious texts, and their Enlightenment experiences often deepen their moral commitment to society. Atheists also feel a deep sense of connectedness—not to God, but to humanity or the entire world. Consider this quote from our survey:

Logic, reason, family, and friends give me all of the wonderful experiences with which I have been a part and desire. I do not desire or expect any supernatural experience, nor do I think that they exist. I consider knowledge and understanding, coupled with compassion and love, to be far more important and satisfying than any religious or spiritual experience.

For those who labeled themselves atheist, there was a consistent pattern of criticism directed toward religion. The reasons for this are complex. Some of our respondents reported bad experiences with religion as a child, even abuse. Others were turned off because they saw hypocrisy between the religious creed and the adherents' behaviors. Some objected to the violence described in sacred texts, or to the centuries of bloody conflict that continues to plague religious groups throughout the world. Rather than surrender themselves to having faith in something they did not experience or believe in, they placed their faith in logic, reason, humanitarianism, and science. Here's an example of how a single criticism caused a child to become a skeptic:

> When I was very young, I attended a funeral. Most of my relatives insisted on telling me that the deceased was going to heaven by being received by angels in the ground. As the casket was lowered, I visualized angels surrounding the person, but when I told my family, they chastised me and told me that it wasn't so. As a result, I assumed that I just made up visions that I wanted badly to see but didn't really exist. It caused me to be skeptical toward every story I hear about religious experiences.

In spite of different backgrounds and different preexisting belief systems, the Enlightenment experiences that both religious and nonreligious people describe are similar. Atheists tend to use the logic-and-reason circuits in their brain to reject religious ideas, while believers use the *same* brain circuits to support their religious beliefs. One of our respondents put it rather bluntly: "My spirituality is based on reasoning and study rather than on a specific experience or event." But notice that he still described reason and study as being something spiritual, merging logic and spirituality into a coherent worldview.

Religious beliefs are often language-based, shaped by the cultures and societies in which we are raised. But Enlightenment transcends words, so it's not surprising that many atheists will also struggle with their old theological concepts. But that's what Enlightenment is about: discovering that the world—and one's belief—is different than what you thought it was. Enlightenment doesn't promise bliss, it offers a deeper truth, and if those truths conflict with older beliefs, internal conflict can result. For an atheist who is married to logic, this can be a problem because Enlightenment defies logic, even though it gives the person an intense sense of clarity.

DRUG-INDUCED ENLIGHTENMENT

One word stood out strongly among those with no religious affiliation or belief: psychedelics. They were the likeliest among the participants to describe drug-induced experiences that brought huge changes in their belief systems. But in our survey, many religious and spiritual people also reported how different drugs stimulated personal transformation. Here's how a sixty-year-old agnostic psychologist described his experience with MDMA, a designer drug commonly called Ecstasy that is purported to stimulate empathy between couples:

> Several years ago, my wife and I were struggling with issues that were threatening the stability of our relationship. We agreed to spend the day on Ecstasy, and we both began to see each other very differently. We resolved issues that had interfered with our intimacy for decades, and the last eight weeks have been some of the best we've ever had. We seemed to forge a new spiritual bond that united us and deepened our love for each other.

From his perspective, there was a permanent change in the relationship that brought about a sense of unity—not with God or the universe, but with each other. Was this a little "e" or big "E" experience? Only the perceiver can make this judgment.

In our survey, many people described their drug-induced experiences as some of the most spiritually significant experiences of their lives:

> The first time I took MDMA, it caused me to decide to go to medical school, become a medical researcher, and at-

tempt to legitimize the use of this drug. During the drug experience, I saw god, but god was clearly a creation of man, not the other way around. Still, I'd have to say that there is a certain something that motivates my actions, a way of being or relating with other people that jibes with the core beliefs of most religions I run across. MDMA opened that door and it profoundly altered me and the way that I exist in the world. I think current religions seem to be groping toward the same goal, largely unsuccessfully.

For Einstein, a compass and a book changed his career. For this experimenter, a psychedelic drug gave him a new purpose, direction, and career. He had clarity, intensity, and a permanent shift in his core beliefs—key elements of Enlightenment.

But there's a problem with MDMA: a recent overview of twenty-five years of empirical research shows serious impairments to memory, sleep, cognition, problem solving, emotional balance, and social intelligence.[7] Other hallucinogens like LSD, DMT, psilocybin, peyote, and ayahuasca can also dramatically alter consciousness and trigger intense mystical experiences leading to positive changes in personality,[8] but they can also have negative side effects that can deeply disturb a person's emotional balance. So the question remains: Can drugs bring Enlightenment? When Dr. Roland Griffiths administered psilocybin to twenty volunteers as part of a research study at Johns Hopkins University, most of the participants reported transformations in their attitudes about life and relationships. The descriptions were generally indistinguishable from the mystical experiences reported in our study: feelings of unity, sacredness, positive mood, transcendence of time and space, and indescribability.[9]

Was there a sense of permanence? Yes! Fourteen months later, Griffiths's participants rated their psilocybin experience as one of the top five most meaningful events in their life. Here are some of the statements they made:[10]

> The experience of death . . . followed by absolute peace.
>
> I was able to comprehend what oneness is.
>
> The breath of God/wind/my breath are all the same.
>
> A more cohesive and comprehensive spiritual landscape became apparent.
>
> The experience expanded my conscious awareness permanently.
>
> The sense of unity was awesome.

In our Western medicine paradigm, we tend to look at drug experiences as artificial because they are "caused" by the drug. But if the drug experience is sufficiently powerful, it might still qualify as enlightenment, perhaps even with a big "E." If, as in the examples given above, a drug-induced state results in substantial increases in compassion, love, and meaning for an individual, what is wrong with that? The problem is this: psychedelics do not guarantee a positive experience. In another study exploring "peak experiences" (Abraham Maslow's term for brief periods of Enlightenment), 47 percent of people who had taken psilocybin reported having transformations of consciousness.[11] But not all of those experiences were positive. One person described "all of the pain and sadness of the world . . . tearing apart my being."

Clearly a "bad trip" can be very disturbing and cause substantial psychological pain. In Griffiths's study, for example, about one quarter of the subjects reported that a significant portion of their session was characterized by anxiety, paranoia, and negative mood. And for people with undiagnosed personality disorders, it can trigger a psychotic state. However, the overall risk of physical and psychological complications is low for most users of hallucinogenic substances, particularly if they are taken in a carefully controlled setting.[12]

Let's go one step further and think about those who specifically take hallucinogenic substances as part of their spiritual tradition. For thousands of years, medicine men and shamans have utilized many psychotropic substances to enter altered realms of consciousness where they report engaging with different spirits, entities, or demons. They use this information to heal others and provide insights into important decisions that would affect the future of their tribe.

I look at it this way. I'm very nearsighted, so when I wake up in the morning, the world is very blurry. Once I put my glasses on, the world becomes crystal clear and I can see things that I would otherwise be blind to without the aid of my glasses. The external reality around me has not changed, but I have transformed my perception of it. What if a hallucinogenic drug is like putting on spiritual glasses for the brain? What if reality has not changed, but our brain's way of seeing it has? Drugs alter brain physiology, and that alteration can lead to a transformational experience, and for a shaman, mind-altering botanicals are the "glasses" they use to tap into a spiritual realm that, in their belief system, truly exists. The research suggests that the brain can be permanently changed by the effects of the drug itself or the transformational experience itself or by a combination of the two.

WILL THE REAL ENLIGHTENMENT PLEASE STEP FORWARD?

When you get right down to it, Enlightenment is the implicit goal of every mystical tradition in the world: to fully surrender oneself to the unity of God, consciousness, truth, or whatever principle is core to the religious or spiritual philosophy. But how many people actually reach these stages of illumination? Few people would probably compare their experience to those of iconic gurus or saints.

Can we ever know who is and isn't Enlightened? And does it even matter? Again, we can only evaluate an intense experience from the inside out. If it changes your life in a way that gives you more purpose and meaning, perhaps it doesn't matter how you label it—big "E," little "e," or a single "aha" moment that illuminates your mind and the world in which you live.

If we aren't aware or don't believe that Enlightenment is a possibility, then we may miss or ignore the subtle neurological changes that allow our brain to see something that lies outside our limited beliefs. To return to Plato's allegory, if we do not know that we are in a "cave" of ignorance, we may never seek a way to escape. But if we know that there is something more awaiting us, then our mind begins to imagine what that bigger reality or truth might be. Perhaps that's why we like to read about enlightened individuals: they give us hope that there is something more wonderful than we currently see or believe. Of course, when it comes to reality, our personal experiences become the ultimate judge.

It is the *belief* in Enlightenment that can start us on the path, and if we recognize its hallmarks—intensity, clarity, unity, and surrendering to a life-changing perspective—then we can consciously engage in activities that will trigger these powerful events. In the next chapter, we'll share with you what we've uncovered about the neuroscience of the "enlightened" brain.

FIVE

The Spectrum of Human Awareness

We are not human beings having a spiritual experience;

we are spiritual beings having a human experience.

Pierre Teilhard de Chardin

From the moment we are born, the human brain has the remarkable ability to constantly change itself. Think about who you are today and who you were a decade ago, or even last year or last month. Although you are the same person, you have learned new skills, had new experiences, and let go of old beliefs and habits that no longer have relevance to your life. We call this process of changing and reaching beyond your current self, "self-transcendence."

Physically, your brain is constantly changing as billions of neurons slowly rearrange their connections in a vast "soup" of neurochemical and neuroelectrical activity. Different electrolytes migrate through the envelope of each neuron, telling it when to rest or take action, and as the activity changes, so do our thoughts and feelings. Throughout the day, different patterns of brain activity

occur, generating different states of mind. Various neurotrans-
mitters are also released when we engage in different tasks, and
each one can alter our behavior and mood. But what happens
when your brain experiences Enlightenment? A growing number
of studies have begun to explore what happens when you have a
sudden insight, and these "aha" moments—as neuroscientists
often refer to them—can help explain how the little "e" experiences of enlightenment influence the brain. Brain-scan research
shows that there is a rapid shift of neural activity in several key
areas of the brain during sudden insights.[1] Our rational mind is
interrupted, our sense of self is altered, and a different form of
conscious awareness emerges. We see problems differently, we
intuitively find solutions in ways that feel surprisingly mysterious,
and in that moment, our knowledge and beliefs can change. So it's
fair to argue that the path to Enlightenment begins the moment
we suspend our normal view of the world.

Some intuitional insights can be intense and bring great clarity, but they do not always radically alter a person's belief system
or create a *permanent* change in behavior or in the function of the
brain, elements that we consider to be the neurological correlates of
Enlightenment. But when we combine the information from the
brain-scan studies of "aha" moments of creative insight with
our brain-scan studies of intense spiritual practices, we can begin
to unlock the biological basis of the big "E" experiences.

HOW YOUR BRAIN GETS YOU TO ENLIGHTENMENT

We have identified five key elements of Enlightenment: a sense of intensity, unity, clarity, and surrender and a permanent large-scale change in our awareness, behavior, or belief system. The little "e" experiences can include any of the first four elements, and the process can be spontaneously triggered (meaning that we are not consciously aware of what brought it about) or deliberately sought through the practices of contemplation, self-reflection, meditation, prayer, or a variety of spiritual disciplines aimed at disrupting our ordinary view of life.

When a person chooses to seek Enlightenment through a specific practice—be it Eastern or Western, religious or secular— activity initially begins to increase in the frontal lobe when she begins to meditate or immerse herself in contemplative reflection. The greater the increase, the more *clarity* we have, allowing us to feel in control and purposeful about our actions and behaviors.

We also see in our brain-scan studies an initial increase in activity in the parietal lobe. Our awareness of our self in relation to the world or object of meditation is increasing, and parietal activity helps us to identify our goal and move toward it.

These initial increases in activity in the frontal and parietal lobes also reduce the emotional intensity of our feelings. This helps us to feel more grounded, centered, and in control, but it doesn't lead to Enlightenment. If, however, there was a sudden and substantial decrease in activity in the frontal and parietal lobes, we would experience a loss of control (surrender), our sense of self would weaken or even disappear, and a dramatic increase of emotion would make the experience feel extraordinarily real.

This is what we've seen in our brain-scan studies. Sometimes when a person is deeply immersed in an intense prayer, meditation, or spiritual practice, there will be a sudden and dramatic decrease in neural activity in the frontal and parietal lobes. This is when our subjects are most likely to describe incredible shifts of perception and experiences of unity consciousness, which are essential parts of the Enlightenment experience. If increased frontal lobe activity helps you to feel in control of your actions, then decreased activity would likely result in a feeling of surrender—another key element of Enlightenment—and if increased parietal lobe activity helps to give you a sense of yourself as a separate entity, then decreased activity would give you the sense that "you" are dissolving and becoming one with everything else in the world, even God. This is where the profound insights associated with Enlightenment begin to enter consciousness.

But how do you distinguish the neurological differences between big "E" and little "e" experiences? "Aha" moments are often very brief, and the increases and decreases in neural activity are rather small. This suggests that the greater the shift in neural activity, the more dramatic the experience will be. When I analyzed the changes in neural activity from all the different spiritual practices we've studied, a certain pattern emerges: the larger the decrease in frontal and parietal activity, especially when they had an initial increase, the more likely the participants were to describe experiences that reflect most of the five elements of Enlightenment.

Let me describe this neurological process with the following metaphor. Imagine that you are slowly climbing two flights of stairs. That's the basic contemplative process during which your frontal and parietal activity slowly increases. You're becoming more conscious, more focused, and more observant of yourself and the world with each step you take. Keep in mind that typically, when you are awake, activity in the brain doesn't change much, perhaps 5 to 10 percent throughout the day. But while the brain is engaged in some of the more esoteric and mystical practices we will be describing in the next section of this book, we see changes approaching 20 percent or more. So clearly you can increase your consciousness and clarity at will.

Now imagine that you quickly run back down the stairs. You'll probably feel a nice rush of energy, like the "runner's high" you get after a good workout, as you come back down to where you started. The same thing usually occurs after you meditate or pray: your brain activity returns to its baseline or resting state. You might feel invigorated and calm, but you probably won't feel Enlightened or transformed.

Then imagine climbing up those twenty feet of stairs—about the height of a diving platform—and jumping off the edge into a swimming pool below. You're descending much more rapidly than running down the stairs, and the experience would feel more intense. That's what happens when your frontal lobe experiences a rapid 20 percent drop in activity. This might resemble a very intense spiritual practice, maybe lasting an hour, and it may even lead you to some new insight or strong feeling of blissfulness—a little "e" experience—but it's not enough yet to trigger a profound shift in your behavior or overall belief system. You had an exhilarating time as you were diving, but your reality didn't change. In this analogy, you are still in the air the entire time—your world has not yet changed.

Now imagine a slightly different scenario where you are standing on a ledge twenty feet above the ocean (the 20 percent increase from your contemplative practice). But then you accidentally slip. Remember, all of the Enlightenment experiences we've reported so far happened unexpectedly to the person, even when they weren't purposely striving for them. You feel yourself rapidly falling to the sea (a 20 percent drop in frontal and parietal activity, which is actually similar to what happens in your brain when you are in a life-threatening situation: your sense of self-control disappears as your survival instincts kick in). Again, though, you are still in the air. You are still in your prior experience of reality.

This all changes when you hit the water. You go from being dry to being wet, from experiencing the world in one way to an entirely different experience. At first, the water doesn't slow your descent as you plummet another twenty feet into the blackness (another 20 percent decrease in neural activity). You are in the depths of the water—an unfamiliar reality—and you can't see anything familiar. This is when we believe you are most likely to experience Enlightenment. Your brain has experienced a 40 percent drop in activity (from the top to the bottom), consciousness is radically changed, and you have no choice but to surrender yourself to the experience. This is how I felt when I was immersed in that sea of Infinite Doubt. I didn't jump. I didn't ask for it. I simply gave up and let the currents take me to the insights that seemed to come from out of the blue.

SPIRITUAL PRACTICES PRIME
THE BRAIN FOR ENLIGHTENMENT

Eastern philosophies placed great emphasis on intense experiences that transcended logic and gave the person a mystical sense of connectedness with the universe. Thus, intense spiritual practice offered the promise of Enlightenment. However, in the West, the Age of Enlightenment was actually anti-enlightenment, at least in any kind of spiritual or supernatural sense. Descartes, for example, would argue that intense feelings of unity and surrender were suspect and needed to be rationally dissected and analyzed in order to arrive at a more simple truth that could be measured by the tools of science.

Mainstream religions in the West were also suspicious of anyone who claimed to experience unity with God, and most mystics were persecuted as heretics. However, the Pentecostal movement that began in the early 1900s, along with the Charismatic movement that followed in the 1960s, changed the course of Christianity. Today over five hundred million people—a quarter of the Christian world[2]—deliberately surrender themselves to becoming one with the Holy Spirit. Our recent brain-scan research with some of these individuals shows that they can take conscious control over their own brain to enter states that meet nearly all of our criteria for Enlightenment.

We studied a group of Pentecostals who engaged in a practice called "speaking in tongues." First, they began to sing and dance as gospel music was played. This was associated with increased frontal lobe activity, but when they began to speak in tongues, activity suddenly dropped in the frontal lobe. They immediately felt an intense sense of unity with something beyond themselves— the Holy Spirit—and as they surrendered themselves to the ecstatic experience, they felt transformed and healed. One of our survey participants described her first experience of speaking in tongues this way:

> I was alone and praying and, without warning, spoke a while in a language I didn't recognize. It was a joyful sound and was a lovely experience and I knew for certain that the Holy Spirit was real, available to me whenever I needed. I couldn't control it, but if I could, I would speak in tongues often!

We also found a drop in frontal lobe activity when we examined Sufi practitioners who engaged in a powerful chanting and movement meditation known as Dhikr. One of our Sufi subjects described the experience of leaving his own body and observing himself from the outside, not unlike the descriptions from people who have near-death experiences.

Many other practices can trigger similar changes, but no matter how it happens, when your frontal lobe activity drops suddenly and significantly, logic and reason shut down. Everyday consciousness is suspended, allowing other brain centers to experience the world in intuitive and creative ways.

Decreases in the parietal lobe activity can also allow a person to have intense feelings of unity consciousness. We saw this in our studies of advanced meditators and people who engaged in various forms of contemplative prayer. But it usually took about fifty to sixty minutes before they felt merged with the object of their contemplation. Our Buddhist subjects described the sensation as becoming one with pure consciousness. The Franciscan nuns in our study felt a sense of unity and connectedness with Jesus or God. These are two entirely different practices, but the unity experience affected the same areas in everyone's brains. This decrease appears to occur even during small moments of insight[3] but is strongest during spiritual practices and powerful Enlightenment experiences.

In our brain-scan studies of various spiritual practitioners, we also saw changes in the thalamus, a central structure that helps us to build reality models of the world. We saw activations in the thalamus during specific practices like prayer and meditation, and we also saw long-term changes in the function of the thalamus in people performing contemplative practices over many years. The more frequently a person engages in meditative self-reflection, the more these reality centers change.[4] Colors can become more vibrant, our empathic feelings toward others can increase, and the way we literally experience the world can become more pleasurable or intense. As one twenty-four-year-old research technician said:

> I once had an experience of extreme light and clarity. My whole body was vibrating for several hours with a light energy that made me see who I was, how I fit in the world, and what fabric was really underlying the physical world in a profoundly new way.

HOW TO VISUALIZE THE ENLIGHTENMENT CIRCUIT IN YOUR BRAIN

Here's a way to help you visualize what happens in the brain when you have an Enlightenment experience. Open up your right hand and imagine placing a walnut on your palm. Now make a fist. The walnut is your thalamus, the part of your brain that receives most of your sensory input and also helps other parts of the brain communicate with one another. Your fist is the emotional center of your brain, and your forearm is your spinal cord. These are the most ancient structures of your brain.

Now take your left hand, spread your fingers apart, and place it over your fist (thumb touching thumb and fingers touching fingers). Your left hand represents your neocortex and its four major lobes. The knuckles of your left hand represent your parietal lobe, which gives you a sense of yourself and different objects in the world, and the palm of your left hand symbolizes your frontal lobe, which controls your conscious decision-making processes. All of these areas of your brain send millions of axons into most of the other areas represented by your fingers, palms, and the walnut inside your fist. In fact, there's so much interconnectivity that you can't precisely say where one part of the brain begins or ends. With each task you perform, you'll see increased activity in some areas and decreased activity in other parts.

Now let's make the image more "real" for your mind to grasp. Take the thumb and forefinger of each of your hands

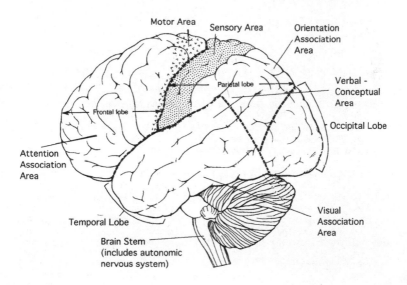

and place them around the circumference of your head. Put your forefingers on the areas right above your eyes, and your thumbs on the back part of your head just above your ears. Your forefingers are touching a very special part of your frontal lobe: your dorsolateral prefrontal cortex. This is the area most active when you are thinking about anything or focusing attention on a specific task like solving a math problem.

Your thumbs are resting on your parietal lobe, where your sense of self originates. Normally, there's a constant dialogue going on between your frontal and parietal lobes. If activity suddenly increases or decreases in either area where your thumbs and forefingers are located, everyday consciousness is radically changed. Your sense of self can expand or contract. You can feel disconnected from reality or unified with it.

THE SPECTRUM OF HUMAN AWARENESS

One of the basic components of the Enlightenment experience, associated with the profound shifts in brain activity, is the radical shift from one state of consciousness to another state, allowing us to see a different view of reality. Consciousness is a difficult concept to describe, but it is essential for understanding Enlightenment. To help with this concept, we have developed a model describing a number of levels of awareness that our brain is capable of generating.

Before describing that model, I want to try to clarify the difference between consciousness and awareness. Many scientists and philosophers use these terms interchangeably, while others attribute unique qualities to each of these subjective states of mind. But when you review the neuroscientific literature, there are actually some important distinctions between them. We are suggesting that awareness is the bigger picture, one that encapsulates many different forms of conscious and unconscious behaviors.

Most biologists would agree that awareness is something that exists in almost every living organism. Even an amoeba moves toward food, suggesting some basic type of awareness of its environment. With the development of the central nervous system in animals, more complex forms of awareness allow animals to voluntarily respond to the environment in more sophisticated ways. This form of awareness involves many parts of the brain.

At some point, the brain develops the capacity to become aware of its own cognitive processes, and that is what we are calling "consciousness," a neurological process that is dependent on—and may be limited to—small areas in the frontal and parietal lobes. When you are *consciously* observing yourself and the world, you become aware that you are a *person,* a unique entity capable of reflecting on your past circumstances and making decisions that will influence your present and future life. You understand, and are aware, that you are you.[5]

Simply put, you can have awareness without consciousness, but you cannot have consciousness without awareness.

Most animals have some awareness of their self and are able

to distinguish that self from other animals and the rest of the world. This is why they won't eat their own leg but rather the leg of another animal. This too is a rudimentary form of self-consciousness. In fact, new brain-scan research shows that many birds or mammals that have structures similar to our frontal and parietal lobes can be consciously self-aware. In fact, dogs have many of the same qualities of consciousness that humans have.[6] This leaves open the possibility that some animals are capable of having "aha" experiences and insights that allow them to alter some of their behavior.[7] However, there is little evidence showing that nonhumans have the ability to radically and permanently change their belief systems, or engage what scientists call "theory of mind"[8]—the ability to understand what another individual is thinking. We also find no evidence showing that animals can consciously alter any structures in the brain associated with the elements of Enlightenment. However, particular animal rituals, like mating rituals, alter the animal's awareness of its own self so that it can connect with another animal. And a similar process occurs in human beings.

Awareness gives birth to consciousness, and consciousness can be enhanced through different practices and rituals. Through these practices, we become more aware of the overall environment and ourselves, including our *unconscious* thoughts, feelings, and perceptions.

But consciousness is also limited by the brain's propensity to rely on old behaviors and beliefs. What if we could "transcend" the limitations of everyday consciousness, as the mystics suggest? Perhaps we can enter higher stages of awareness that allow us to glimpse greater truths about reality. Is it possible for people to tap into new levels of sensory experience that would expand their awareness of the world? Perhaps! That is why we are proposing a neurological model to include the possibility of Enlightenment.

Based on our analysis of the last twenty years of neuroscientific research on learning, memory, emotions, cognition, and behavior, along with our own studies of spiritual experiences, we have identified six "levels" of awareness:

Level 1: Instinctual Awareness

Level 2: Habitual Responsiveness

Level 3: Intentional Decision Making

Level 4: Creative Imagination

Level 5: Self-Reflective Awareness

Level 6: Transformational Awareness

Each level is associated with activity in different regions of the brain, as the following diagram shows. With this map, we can now trace how awareness emerges in the primitive nervous system of a worm[9] and culminates in the Enlightenment experience of a human.

ENLIGHTENMENT

SPIRITUAL — INTUITIVE
PSYCHOLOGICAL — SUBJECTIVE
BIOLOGICAL — PERCEPTUAL

6 TRANSFORMATIONAL AWARENESS
(life-changing insights that provide deeper meaning and purpose)
structual changes in thalamus, neocortex, midbrain, amygdala

5 SELF-REFLECTIVE AWARENESS
(intuitive, nonjudgmental, and symbolic processing of self and others)
anterior cingulate, insula, precuneus, enhanced frontal/parietal activity

4 CREATIVE IMAGINATION
(mind-wandering, daydreaming, and free-associative fantasies)
resting-state network: interhemispheric activity, memory reconsolidation

3 INTENTIONAL DECISION-MAKING
(language-&-feeling-based processing, conscious goal-seeking)
dorsolateral pfc & surrounding areas, temporal/parietal areas

2 HABITUAL RESPONSIVENESS
(learning, memory formation and recall, unconscious behavior)
posterior neocortex, midbrain, cerebellum

1 INSTINCTUAL AWARENESS
(sensory and emotional processing; survival-based motivation)
thalamus, nucleus accumbens, ofc, PAG, amygdala

A BRIEF SUMMARY OF THE LEVELS
OF HUMAN AWARENESS

Our brain often engages several levels of consciousness and awareness at the same time, but we spend most of that time in the lower levels of instinct, habit, and conscious decision making. For example, you wake up in the morning feeling hungry (Level 1, instinct). You automatically put on your robe and walk to the kitchen, barely aware of anything (Level 2, habit). You open the fridge and decide what you'd like to eat (Level 3, decision making). That's about it. Maybe you'll be inspired to make something new for breakfast (Level 4, creativity), but I doubt that you'll spend much time deeply reflecting on the meaning and purpose of your daily routine for getting up (Level 5, self-reflective awareness). Your beliefs are the same, your behavior is predictable, and your worldview will not have changed (Level 6, transformational awareness).

If you are like most busy people, you probably gobbled down your food and rushed out the door, hoping that you wouldn't be late for work. Not much awareness there, but what if you engaged in a practice called mindful eating, where you savored each tiny bite of your food? We recently did a brain-scan study on mindful eating and discovered that there are some very unique neurological changes that take place. When you eat super slowly, food tastes surprisingly different. For one of our test subjects, desserts began to taste slightly unpleasant, and the experience was so shocking that he eliminated sweets from his diet. Eating slowly made him more self-aware (Level 5), and the change in behavior probably lengthened his life (he was pre-diabetic). We would consider this a little "e" experience because it had the elements of surrender (he immersed himself in the experience of eating slowly), intensity (the flavors were accentuated), and clarity (he realized he didn't like what he normally ate), and it changed his behavior (his beliefs about what tasted good were challenged by the mindful eating experience).

With Enlightenment comes increased awareness, but this example demonstrates that even during ordinary activities, we can deliberately alter our awareness to experience the world in very different and beneficial ways. In other words, increased awareness is a key path to Enlightenment.

Enlightenment requires that we take time to access our intuitive centers of creativity and awareness and to deeply reflect on those things that give our lives meaning, purpose, and value (Levels 4 and 5). The more you understand how to shift between the different levels of the Spectrum of Human Awareness, the more you are likely to experience Enlightenment.

Let's explore the different levels in more detail.

LEVEL 1: INSTINCTUAL AWARENESS.

It begins the moment we wake up and voluntarily respond to our inner and outer needs. This basic level of human awareness is mostly unconscious and is governed by our emotions and our pain-versus-pleasure reactions. It's survival oriented, directing us to move toward enticing goals and away from potential threats.

LEVEL 2: HABITUAL RESPONSIVENESS.

As we move through life's challenges, we develop new skills, embedding them into long-term memory. Slowly, over many years of childhood development, we build a repertoire of habitual behaviors that we use to achieve most of our goals. Usually we are unconscious of our habitual actions, but we become aware of them when they interfere with our goals.

LEVEL 3: INTENTIONAL DECISION MAKING.

This level of awareness is where most of our logic, reason, and attention go to solve simple problems and accomplish day-to-day tasks. In our previous book *How God Changes Your Brain*, we referred to it as our "everyday" consciousness to distinguish it from the self-reflective awareness that leads to personal transformation. Everyday consciousness is also very limited, and when we are working on a task, we aren't very aware of much else that is happening around us. Everyday consciousness is related to our short-term working memory, which contains only enough information to make moment-to-moment decisions. These processes take place in specific areas of our frontal lobe.

Here's a simple demonstration of the limited awareness that exists on Level 3: You are probably very aware of the words you are reading right now, but not the chair you are sitting in or most of the ambient sounds in the room. However, the moment you consciously shift your attention to the chair and the sounds, you'll find that you can't continue reading. That's how everyday consciousness works, but we rarely notice how limited it actually is.

In fact, when we are making decisions and moving toward specific goals, we are only barely aware of what is happening in the present moment.

LEVEL 4: CREATIVE IMAGINATION.

Decision making and goal achieving is a strenuous neurological process, and the brain must take frequent breaks in order to reset its neurochemistry. Normally, this process involves the relaxation of both our body and our thought processes, and if we fail to heed the signs of fatigue, we can experience severe emotional stress.

It is essential to take hourly relaxation breaks from work. As you relax, some areas of your brain actually become more active, especially in your frontal lobe, and your mind begins to wander and daydream. Not only is your brain refreshing the neurochemicals needed to make decisions, it is actually engaged in creatively solving problems.[10] It's also essential for memory formation and recall, but most important, this relaxed state of daydreaming allows you to enter the higher stages of awareness where the path towards Enlightenment often begins.

Creative imagination can also occur when a person purposefully focuses awareness on a given problem and finds a new kind of solution. This process of creative imagination may utilize the different sides of the brain to posit a problem and come up with a more holistic or artistic solution. Many scientific and mathematical discoveries are made this way.

LEVEL 5: SELF-REFLECTIVE AWARENESS.

Most people unconsciously slip in and out of the relaxed mind-wandering states of Level 4, never realizing how useful they can be. But if you purposely choose to suspend the everyday consciousness of Level 3 by relaxing into the imaginative processes of creativity, you begin to broaden your experience of the other levels of awareness. You can identify unconscious habits—good or bad—and you can begin to see the instinctual drives that motivate your brain to take action in the world. This gives you a greater ability to evaluate problems and make better decisions. When you consciously daydream and mindfully reflect on all the seemingly chaotic thoughts and feelings that float in and out of consciousness, brain scans show increased activity in the left prefrontal cortex (where your sense of clarity and optimism resides) and decreased activity in the right prefrontal cortex (which tends to generate and process worries, fears, and doubts about future actions).[11]

In fact, most forms of negative thinking interrupt the brain's ability to perform well on every level of the Spectrum of Human Awareness. Self-reflective awareness—often called mindfulness—is one of the few documented strategies that will make you aware of your different levels of consciousness. As our brain-scan studies have shown, conscious engagement of Level 5 causes substantial increases of activity in the anterior cingulate cortex and insula, areas that help regulate emotions while increasing the brain's ability to empathetically connect with others. You are turning on more awareness centers of the brain than you typically use when you are operating in the lower levels of the Spectrum of Human Awareness, literally expanding consciousness.

The simple act of watching your own consciousness actually improves your mood, your self-esteem, and your overall satisfaction with life. Research also shows that self-reflective observation and awareness activates structures in the brain directly associated with Enlightenment and transformation.[12]

As you observe your creative mind, you will become aware that "you" are not your thoughts. This for many people is a profound insight and is directly related to the Zen Buddhist concept of Enlightenment. Negative feelings lose their power as a sense of inner serenity engulfs you, and while this happens, you turn on thousands of stress-reducing and immune-enhancing genes.[13] You are neurologically transforming the structure and functioning of your brain, and this is what allows you to enter the highest level of our Spectrum of Human Awareness.

LEVEL 6: TRANSFORMATIONAL AWARENESS.

Here we radically diverge from the previous levels and the research supporting them, and for good reason: How can you *document* that you've actually been Enlightened, and that it's not just another fantasy generated by your imagination? It's a difficult task, which is why we recommend that you keep an open mind when experiencing moments of profound, life-changing insight. However, several research teams have been able to identify those periods of exceptional thinking, where one's perspective suddenly changes and worries and anxieties miraculously melt away.[14] They involve many of the same brain areas we've identified in our studies,[15] and this is why we believe that transformational awareness is a *subjectively and neurologically* real experience that causes people to achieve Enlightenment. These individuals consistently report more happiness and satisfaction, and the neurological changes we see suggest that Levels 5 and 6 can even slow down the aging processes in the brain.[16]

The moment a person first experiences transformational awareness, he or she might feel that Enlightenment has occurred. It can last for seconds, minutes, hours, or even days. But at some point, the brain will return to its habitual way of functioning and decision making. However, it is not the same brain! Our previous studies document that subtle and permanent changes will have taken place in key brain areas. Thus, the everyday consciousness we return to is not the same consciousness we had before. We've changed. We have new knowledge about ourselves and the world and we have increased our ability to become more fully satisfied with our life.

The Spectrum of Human Awareness tells us that we can take specific steps to consciously improve our lives, and it shows us what we need to do if we want to actively pursue Enlightenment. As psychologists at Drexel University emphasize, you can prepare your mind ahead of time to encourage brain-changing activity associated with sudden insights.[17] It will make it easier for you to reach Level 6, which opens the door to the incredible personal transformation of Enlightenment.

PART 2

THE PATHS TOWARD ENLIGHTENMENT

●

To see a world in a grain of sand,

And a heaven in a wild flower,

Hold infinity in the palm of your hand,

And eternity in an hour . . .

God appears, and God is light,

To those poor souls who dwell in night.[1]

—*William Blake, eighteenth-century English poet*

SIX

Channeling Supernatural Entities

In the next few chapters, we are going to share with you some of the most recent and fascinating brain scans we've done, and they all share one thing: activity in the frontal or parietal lobes rapidly decreases. This unusual neural activity, as we described in the previous chapter, is associated with radical shifts of consciousness, something that appears to occur when a person experiences Enlightenment. These studies are also unique since most forms of contemplative practice (mindfulness, Buddhist meditation, Christian Centering Prayer, etc.) show a gradual *increase* in frontal and parietal activity.

When frontal lobe activity decreases, however, normal states of consciousness and communication are disrupted. And when activity in the parietal lobe decreases, a person's sense of self seems to disappear, creating a neurological sense of "unity." Practitioners of more unusual or esoteric practices (which include channeling, chanting, conveying divine wisdom to another, etc.) often feel that there is no separation between themselves and God or the forces that shape the universe. Everything seems interconnected as one's sense of self begins to dissolve. In this rarified state, sensory awareness is heightened and the practitioner begins to experience reality in a different way.

CONNECTING TO THE HOLY SPIRIT

I first became interested in more esoteric spiritual practices when, in 2005, *National Geographic* asked me if I would be willing to do a brain-scan study with people who felt they were possessed by demons. Feeling somewhat uncomfortable with the topic, I proposed an alternative: to study the Pentecostal practice of speaking in tongues, a historical form of divination that can be traced to ancient Christianity and Hasidic Jewish mysticism.[1]

Members of a Pentecostal church will begin to sing and sway as they listen to a sermon or Gospel music. Then they invite the Holy Spirit to enter their consciousness. Our research showed that when this occurs, practitioners enter a trance state—usually in just a couple of minutes—and many begin to speak in a manner that sounds like a strange foreign language.[2] They feel that the experience enhances their spiritual connection to the divine, and many Pentecostal churches equate this experience as a direct form of Enlightenment bestowed by the Holy Spirit. In fact, the Pentecostal movement of the early 1900s gave birth to the notion of "evangelical enlightenment," where direct encounters with God would help to transform the consciousness of humanity, bringing universal liberty, equality, and prosperity to everyone.[3]

When our Pentecostal participants began to speak in tongues, their frontal lobe activity immediately declined. The language areas in the frontal lobe also decreased. Normally, when you talk and listen to others, the communication centers in your brain turn on, but when a person begins speaking in tongues, these areas shut down. Since some type of vocalization is occurring, either God is actually taking over (as the Pentecostal would argue) or perhaps an alternative type of communication process is being activated. Either way, the sense that these vocalizations are coming from another realm would feel very real. And as all of our subjects reported, the experience made them feel ecstatic, likely related to a strong activation of the emotional or reward centers of the brain.

People who speak in tongues find the experience incredibly powerful and transformative. They gain many new insights into how to lead their life—a little "e" enlightenment—and sometimes their experiences bring them to their knees with tears streaming down their face. At these moments, most people will experience a radical shift in their emotions and thoughts—a big Enlightenment. They feel connected to God in a way they never felt before and this is accompanied by great clarity and realness. Often, it will lead a person's life in a new direction.

Similar shifts of consciousness, and brain activity, occur during other forms of trance states, such as when mediums try to contact the dead or when Shamanic healers enter the spirit realm. We believe that these dramatic neurological changes are important paths that can lead to Enlightenment experiences.[4]

WHAT IS A TRANCE?

The typical dictionary definition describes a trance as a sleep-like state where a person is barely conscious, experiencing diminished cognitive and body activity. To the outside observer, the person could appear to be in a coma, but for those who deliberately choose to enter a trance, the experience is very vivid. In a trance, you are barely aware of yourself, but depending on your intention, you can trigger a variety of visions or sounds that have a hyper-real or supernatural quality. In this altered state of consciousness—what we would consider a very intense form of creative mind wandering (Level 4 on the Spectrum of Human Awareness), practitioners seek information or insights that are not accessible through normal forms of conscious activity (Level 3).

Documented trance states date back at least to ancient Greek temples where divinely inspired people would give advice to others or predict future events. Many saints from the Middle Ages would also enter trance states to feel connected to God, and they would use strategies like gazing at objects or paintings, chanting repetitive phrases, or doing ritualistic body movements to interrupt everyday consciousness.

WISDOM FROM THE DEAD?

Can trance states, which supposedly connect a person to the spirits of the dead, provide a path toward Enlightenment, or at least shed some light on what Enlightenment is? We think so. A few years ago, a Brazilian friend of mine, Julio Peres, who is a researcher in psychology, began studying a group of highly educated mediums who practice a technique called psychography. To perform psychography, the medium enters into a trance state and connects to the spirits of the dead. The spirits provide information, which the medium automatically writes down.

During these sessions, the mediums do not feel like they are guiding their own hands. Rather, it is the spirits who are causing their hands to move. The mediums then share their information with friends and family members of the deceased. Interestingly, a triple-blind study conducted at the University of Arizona showed that the mediums they tested were more likely to give accurate information about the deceased than those relatives who were not mediums.[5] The mediums were able to identify more aspects of the deceased relative in terms of physical appearance, personality, hobbies, and the cause of death.

For over a hundred years, mediumship has been very popular in Brazil, and I was very interested to see what was occurring in the brains of these psychographers. Would their scans look like activities we found in meditators, or Pentecostals, or some completely new pattern? I was also interested to know how the practice of being a medium affected their lives.

Mediumship can be traced back to the Book of Samuel in the Jewish bible. However, many critics view most mediums as charlatans, especially since so many have been exposed as frauds by various law-enforcement groups. Other critics consider the practitioners delusional, but the research disagrees. For example, a recent study of 115 spirit mediums found that they had a low prevalence of psychiatric disorders and were well adjusted socially compared with the general population.[6] Clearly, their trance states were distinct from dissociative identity disorder, and many psychologists now recognize that such practices can be part of a legitimate spiritual practice.

Other researchers have compared the trance states of mediums to self-induced hypnotic techniques that allow a person to access subconscious thoughts or feelings.[7] Hypnosis, however, involves increased activity in the frontal lobe and is accompanied by a perceived sense of increased awareness.[8] In the trance states I have studied, we actually see a *decrease* in frontal activity, associated with a partial loss of consciousness.

The psychographers reminded me of the famous psychic Edgar Cayce, who by 1912 was world renowned for his ability to enter trance states to gain knowledge about a person's illness and what remedies could be used. I wondered what types of brain changes might be happening in his brain. People called him the "sleeping prophet," and when Cayce would awaken from his trances, he'd have no memory of what he said. But while he was in trance, a strange voice would emanate from him, which his assistants would record. This is similar to the Pentecostal practice of speaking in tongues, where frontal lobe activity also decreased. Setting aside the question of actual psychic ability, trance states deserve special attention because they appear to measurably enhance spirituality and life satisfaction in people who consciously choose to elicit them. For this reason, I created a research protocol to shed more light on the practice of mediumship. What we found was an incredibly healthy group of individuals who provided a valued service to their community. We published our brain-scan findings in a peer-reviewed journal in 2012.[9]

INSIDE THE BRAIN OF A MEDIUM

Neuroscience cannot truly test whether a psychic or a medium is actually connecting to the spirit world, but we can explore what happens in the brain when the medium enters these unusual states of consciousness. We can then compare those changes to other spiritual practices and also identify potential health benefits.

We devised an experiment using single photon emission computed tomography (SPECT) to measure different regions of the brain. When certain areas become more active, there is increased blood flow, and if that occurs in the frontal lobe, for instance, your decision-making skills increase. If it occurs in the parietal lobe, your conscious awareness of yourself may increase. If it occurs in the amygdala, you might feel suddenly fearful, and if it occurs in the thalamus, we believe that the event you are experiencing will feel more real and intense.

To do a SPECT scan we start by placing a small intravenous

catheter in your arm. Then, while you are performing a particular activity—in this case, entering a trance state—we inject a small amount of a radioactive tracer that quickly travels to the most active areas in your brain. These tracers are generally considered quite harmless since the several nanograms of material are so small. Importantly, once the tracer gets to the active part of the brain, it stays there. So after you've completed the activity (for example, prayer or psychography) we want to measure (the trance state in this case), we'll take you down the hall to our SPECT camera and literally take a picture of what your brain was doing at that moment.

For our trance study, my friend flew up a group of expert practitioners of psychography from Brazil. When I met these mediums, they were extremely gracious and pleasant. They were very excited about being part of the study, and several of the mediums said that they had already received messages from the spirit world encouraging their participation.

Right before we started, one of the senior members of the group commented that there was a male spirit who was near my head and who was a family member—an uncle on my father's side of the family. I was told that the name of the spirit was Joseph and that this spirit was very interested in the research I was doing. When I returned home that day, I immediately called my father to ask him if we had any uncles named Joseph who had died. We did!

Now Joseph is a very common name, but the notion that *maybe* the medium actually contacted someone excited me. I thought that if great-uncle Joe was interested in my research on spiritual experiences, maybe he was a particularly spiritual person, perhaps a rabbi. That's where my imagination took me, but when my father said that Joseph was an accountant, I felt disappointed. After all, why would an accountant care about a brain study of this type? Maybe he was a particularly spiritual accountant. We may never know, but I share this story because many research studies show that there are parts of the brain that have no problem believing in supernatural powers.

WHEN THE SPIRIT MOVES YOU

For our study design, we decided to compare the brain activity during the practice of psychography to that of normal writing. Had we compared psychography to a simple resting state, we certainly would have observed a variety of changes going on in the brain. But we wouldn't know if the changes were related to writing, moving the hands, looking at the page, daydreaming, or actually performing psychography. We wanted to find out if psychography was a different kind of writing than normal writing.

For all participants, we placed a small intravenous catheter in their non-writing arm. We asked them to write about an everyday event, about their interest in this research study, or anything creative that came to mind. After writing for ten minutes—which they did quite furiously, by the way—I injected them with the radioactive tracer, and about five minutes later we brought them to the scanner to take a picture of their brains' activity. We saw activity in the temporal and frontal lobes, which we know to be part of the language center. We expected this type of pattern because it is similar to the activity seen when people engage in normal writing and reading tasks.

When the first scans were finished, we brought the mediums back into the room, where they began their psychography practice. They started with a brief meditation, then a prayer, and then they sat quietly, pen in hand, waiting for the spirits to tell them what to write. After a few minutes, they began again to write furiously. As an outside observer, I really could not tell the difference between the normal writing state and the psychography state. They just looked like they were writing words on a page. Ten minutes passed and I injected them with another dose of the tracer. They continued to perform their psychography for another five minutes, and when they had finished, we placed them in the scanner to get a picture of their brain's activity during their trance state.

The expert mediums showed a dramatic decrease in frontal lobe activity, similar to what we observed in the Pentecostals when they spoke in tongues. The mediums also showed dramatic decreases of activity in the temporal lobe, particularly in the areas involved with language. It's as if their communication center was actually being taken off-line during the psychography practice. Clearly their brain was not writing the way it normally would, and yet the writing was lucid and precise.

During psychography, intentional consciousness (Level 3) was interrupted. But if the mediums were entering a creative intuitive state of awareness (Level 4), we would expect to see *increased* activity throughout the frontal lobe. We didn't. Here's where our

Spectrum model is useful. We know, for example, that habitual behaviors (Level 2) require very little frontal lobe activity, and so we suggest that the mediums are writing from memory. But Level 1 awareness is always operating whenever you are awake. This is where the brain is emotionally responding to whatever is happening in the present moment. Whatever information is being perceived by the mediums would feel like it is actually coming from the outside world. Instead of purposefully forming words, as we might do in a normal conversation or when writing a letter to a friend, these mediums tap into a unique state where language appears to flow from a different source.

You typically have to train yourself to decrease frontal lobe activity, and thus we see two different strategies: speaking in tongues and channeling spirits. In the following chapters, we'll show you other strategies, but the bigger question is this: What *value* does it serve to obliterate everyday consciousness? Normally, the thoughts that fill our mind make it difficult to be cognizant of the other levels of awareness we outlined on our Spectrum model. For example, when we are focused on a daily task (eating, brushing our teeth, driving to work, meeting a deadline), we are often oblivious to the subtle sensations generated by our emotional responses to the world (Level 1: instinctual awareness). We don't notice the discrete pleasures occurring in our body while we move, or the tiny aches and pains caused by muscle tension. Even feelings of love, caring, or curiosity can be hidden behind out goal-and-worry-oriented thoughts.

When we interrupt everyday consciousness, we also interrupt

our habitual patterns of thinking and behavior (Level 2), and this allows us to experience more of the world around us. In this altered state, it becomes easier to see what really motivates us, and what our basic human emotions feel like (Level 5).

Getting outside your usual level of consciousness is what appears to move a person toward both little and big enlightenment experiences. In the communities in which these mediums live, their practice helps them understand the world in new ways, and it also helps many others find new meaning in life, qualities that reflect the Enlightenment experiences we've described in the earlier chapters of this book.

What makes these experiences so different from contemplative meditation is the time factor. In our studies of Franciscan nuns and Buddhist meditators, it takes about fifty to sixty minutes to create these same kind of neurological changes. The Pentecostals and mediums took far less time, sometimes only minutes, to enter such altered states. This suggests that it may be very easy to prep the brain for transformation simply by picking up a pen and asking for advice from an entity or sage, living or dead, imaginary or real. By giving up habitual control, we may gain access to deeper wisdom within and beyond the boundary of everyday consciousness.

"FLOWING" TOWARD ENLIGHTENMENT

In our study, some of the mediums were beginners, and we did not see a decrease in the frontal lobe. This suggests that it takes time and practice to dramatically change neural functioning in a way to deliberately trigger an experience resembling Enlightenment. As with anything we wish to excel at, there's a learning curve. What, then, is the beginner's brain doing differently?

To answer this question I would like to use a musical analogy. Many years ago, when I was a child, my parents had me take piano lessons. At the beginning, it was painfully difficult to manipulate the keys in a way that sounded like music. I had to concentrate very hard on each finger, each key, and each note that emanated from the piano strings.

After a few months, I could play a few simple tunes, even a few from famous composers, but I still had to concentrate on every note. It was music, but it certainly didn't sound like Bach. The more I practiced, the better it sounded, and I didn't have to concentrate as much. Concert pianists, through years of practice, reach a level of skill where something else takes over in the brain—where concentration is replaced by intuition, improvisation, and finally inspiration. At that moment, they alter the normal functioning in their frontal lobe,[10] shifting from the everyday consciousness of Level 3 to either a more instinctual form of playing or a more creative form of expression. When that occurs—when old habits and intentional decision making is interrupted, the music transforms itself from excellence to euphoria, or what Mihaly Csikszentmihalyi calls a state of "flow," a well-documented psychological state where you fully immerse yourself in an activity. You lose awareness of everything else, even feelings of hunger or tiredness, as your sense of self dissolves and you become filled with a sense of unity and happiness. Of course, the activity must be challenging *and* enjoyable,[11] and you must also practice regularly if you want to bring this state of mind into your life.[12] We, and many others, believe that flow is an essential ingredient for reaching any state of enlightenment.

Musicians who reach flow often feel as if their hands are playing the piano by themselves, not unlike the way the Brazilian mediums described their automatic writing. The more you practice letting go of your normal consciousness, the more your brain changes in ways that allow you to feel connected to the experience itself.

WRITE YOUR WAY TO ENLIGHTENMENT

Another interesting finding arose from our research on automatic writing. My Brazilian colleagues analyzed what was actually written during the psychography practice and during normal writing. The results showed that the written content was much more complex during psychography. This is fascinating because you would think that more complex writing would require more activity in the usual language areas. But somehow the experienced mediums were able to produce more richness and variation—much like how a great poet composes a line of verse—even without the usual language areas. So maybe there really is some type of external "communication" coming to these people. Of course, even if this were the case, it does not mean that it came from dead spirits. Perhaps they are simply picking up on information coming from other people around them.

While psychography alters consciousness and affects the brain in unique ways, we would not say that automatic writing, in and of itself, was a form of enlightenment, especially since the practice did not change the mediums' belief system. Perhaps their first experience may have changed their life and started them down a new path, but it is possible that the practice of automatic writing can help to open up our consciousness to new ways of thinking, ultimately leading to Enlightenment. In other words, it may prime the pump through little "e" experiences.

We have tried to learn from our studies what types of practices can produce the greatest changes in the brain, specifically the big drops in frontal lobe activity that interrupt habitual states of consciousness and lead to a sense of surrendering one's self.

Practices like psychography or speaking in tongues can stimulate unique circuits in the brain that give one the power to "channel" information from what feels like different entities or sources of inspiration. Whether such experiences are purely created by brain activity or actually come from a spiritual dimension, science cannot say. But the research is definitive: *any experience*, if it brings enough clarity to change our behavior or beliefs, can lead to a little "e" or big "E" moment of insight.

If you would like to experience automatic writing right now, go grab a pen and some paper and try this experiment.

STEP 1:

First, write down the first thought that comes to mind—anything. Congratulations! You just wrote from your "normal" state of consciousness (Level 3 in our Spectrum model).

STEP 2:

Now take a moment to recall a time when you were filled with anger or rage. Fill yourself with that feeling and then write down whatever comes to mind, using whatever expletives you'd like. Now think about something that made you feel very sad. Immerse yourself in that feeling and write down the words that automatically come to mind. You've just accessed, the best one can, the first level of instinctual emotional awareness described in our Spectrum of Human Awareness.

STEP 3:

Now for the fun part: write down the craziest sentence you can think of. Imagine that you are a lunatic, or stoned, or drunk, and write down something ridiculous (for example, there's an elephant wearing a diaper sitting in my refrigerator). Keep writing down crazy, wild, silly sentences until you feel a sense of abandonment. It doesn't matter if you scribble gibberish or make up meaningless words. You are now exercising Level 4, creative imagination. This, by the way, is a common warm-up exercise used by many writers to break through writer's block.

STEP 4:

Take ten very deep breaths, or run in place as fast as you can for thirty seconds. This is a quick way to interrupt everyday consciousness and prepare your mind for automatic writing. Next, just sit quietly for a moment and watch your thoughts and feelings ramble through your conscious mind. This is the first stage of

Level 5, self-reflective awareness. Now think about an issue or problem you are currently struggling with, and imagine that you are someone else—the world's greatest problem solver, like Freud or Einstein, or Harry Potter—and begin to write a response to your problem as if you were that other person. Write down whatever comes to mind without censorship. Trust your intuition and be as inventive (or silly) as you like in one or two paragraphs. Then just gaze at what you wrote, staying deeply relaxed and observant, and ask your intuition to provide a new insight into your problem. You'll often be surprised as you receive a glimmer of inner wisdom that seems to come from nowhere. That's the beginning of Level 6, where personal transformation begins.

After a bit of practice, you may discover that automatic writing can help you solve problems better than normal journaling can!

One study found that doing a creative writing exercise like this one was particularly effective for processing feelings of grief.[13] Another study, conducted at Carnegie Mellon University, found that creative writing can interrupt negative thoughts and depressive symptoms, whereas "normal" writing had no such effect.[14] Interestingly, as with the mediums, researchers found decreases in the language centers of expert creative writers.[15]

So if you really want to pursue a path toward Enlightenment, the answer is simple: keep practicing!

SEVEN

Changing the Consciousness of Others

When people have profound experiences that change their life and worldview, they usually want to share their insights and wisdom with others. Even if the experience itself is difficult to put into words, history is filled with the teachings of "enlightened" individuals. Even in Plato's allegory of the cave, the message is clear: if you have seen the "light"—if you have ventured from the darkness of ignorance and now see reality for what it really is— you must return to your companions in the cave and pass your knowledge on to them.

In this chapter we're going to explore one of the most esoteric forms of communication that relates to the power of our silent thoughts to influence others. In the East, the ability to silently pass wisdom from the master to the student was called "dharma transmission," which could result in sudden Enlightenment. The concept was popularized in the seventeenth century when the Buddhist monasteries and traditions were under severe political attack, and it continued to be practiced as a way to identify the lineage of revered teachers and gurus.[1] Today the practice of enlightening others just through the gaze of the master is sometimes called "shaktipat."

CAN THOUGHTS HEAL?

In the West, the notion that one's thoughts could influence others was a foreign concept often associated with witchcraft, sorcery, or brainwashing. When it came to spiritual enlightenment, most Western religions believed that God was the only bestower of such wisdom. However, one could petition God with one's thoughts through prayer, and if you wanted someone to be healed, you had to go through God, or the head of the religious organization.

All of that changed in the late 1800s, thanks to the influence of Mary Baker Eddy and her Christian Science movement. Eddy had been a patient of Phineas Quimby, a man who believed that all illness was caused by the mind, and she began to integrate a wide variety of metaphysical healing strategies with biblical passages. The result: many people came to believe that they could directly tap into God's divine power to heal others who were suffering or ill.

According to mainstream Christianity, this was almost heretical, because God was taken out of the equation. However, the Christian Scientists believed that everyone could be enlightened by the Holy Spirit or Holy Ghost and directly transmit healing powers to others. Mental healing became the new form of prayer and it rapidly spread through America, giving birth to dozens of churches who viewed God, consciousness, the mind, and the universe as one and the same. Eastern and Western notions of Enlightenment were finally united, and the belief that thoughts could heal other people at a distance now has hundreds of millions of believers throughout the world. But can silent thoughts really heal? Scientific studies on intercessory prayer—the ability to affect another person's health from a distance—have been conducted for many years. When you look at the early studies, the evidence was positive: people in hospitals who were being anonymously prayed for were healing a few days faster and being released from the hospital a few days earlier,[2] and it was shown to help coronary care patients and patients suffering from arthritis.[3]

But beginning in 2001, as more studies were conducted around the world, the benefit claims began to disappear.[4] Intercessory prayer had no effect on those who suffered from alcohol abuse,[5] child psychiatric disorders,[6] and immune disorders[7] or on the healing of wounds[8] or the health of pregnant women.[9] Nor was there any effect on the growth of cultured human brain cells, although the study did show that a mechanical random-number generator was affected.[10]

Some studies even showed that those who knew they were being prayed for had more problems and complications than those who were not prayed over.[11] Another study found that those who believed in prayer did better, and those who didn't believe showed no improvement of symptoms.[12] Still other researchers suggested that when active prayer was integrated into the doctor-patient relationships, it could "strengthen the patient's optimism and activate the body's healing resources."[13] In countries like Brazil and Puerto Rico, where mediums are an accepted part of the treatment in mental health facilities, the patients report many benefits.[14]

What is going on? Do our healing thoughts have any effect on others, or worse, can our prayers actually cause harm? These are important scientific, medical, and ethical issues to address. Certainly a person's belief system plays a major role in these experiments. For example, in 2014, when Ankara University researchers in Turkey measured the therapeutic effects of Islamic intercessory prayer on Muslims infected with warts, they found significant improvements in patients who trusted the person sending the prayers, but saw no effect at all in patients who did not.[15]

Here's what the research shows: if you believe that prayer heals, that alone can stimulate the immune system. But if you don't know if you are being prayed for, there is no convincing evidence showing that thoughts at a distance have an effect.

A recent brain-imaging study made a remarkable discovery. A group of subjects—half of which were devoted Christians and half who were secular—was studied with fMRI while listening to a series of recorded prayers. The subjects were told that the prayers being said were by either a non-Christian, an "ordinary" Christian, or a Christian with healing powers. However, the prayers were all being said by the same male speaker, a Christian who did not belong to any charismatic group. When the devoted Christian subjects (who were mainly Pentecostal) listened to the prayers they believed were coming from a Christian endowed with healing powers, activity in their frontal lobe decreased.[16] Based on our trance-state research and our neurological model of enlighten-ment, this suggests that believers can enter a state of conscious-ness where healing powers feel utterly real. However, our research does not show that healing prayers stimulate any form of insights that could lead to enlightenment experiences. In a research study I recently conducted—and which I'll talk about in more detail in a moment—I tested a distinguished healer's ability to influence a person's brain when she was in an adjoining room. As far as we could see, nothing happened. When I scanned the healer's brain, I found changes similar to other meditation studies: frontal lobe activity went up, suggesting that there were physical and mental benefits for the healer—but not for the person being prayed for. This strongly suggests you must deliberately seek Enlightenment for yourself by actively engaging in processes that will interrupt your normal way of thinking and experiencing reality.

PARANORMAL EVENTS *REALLY* HAPPEN!

In religious literature from around the world, many enlightened people are said to have gained special powers as a result of their personal transformation. In our online survey, we asked our re-spondents if they believed they had developed any unusual capa-bilities as the result of their spiritual epiphanies, and 10 percent claimed to experience telepathy, telekinesis, clairvoyance, or the ability to communicate with the spirit world. But does Enlighten-ment actually endow people with special powers, and can we even study whether or not supernatural abilities truly exist?

Parapsychology is the study of how thoughts, prayers, consciousness, or human "energy" can influence other people or objects. It also includes the study of psychic (or psi) phenomena such as telepathy, precognition, clairvoyance, near-death experiences, reincarnation, mediumship, and other paranormal claims.

In spite of the generally negative view that most scientists have on such research, there are a number of well-designed studies showing consistently positive results that people can actually affect other objects at a distance. Although the effects are very small—often less than one percent—these studies indicate that the results are better than chance. For example, researchers at the University of Missouri analyzed the data from hundreds of psi phenomena studies and came to this conclusion: "We find that the evidence . . . favors the existence of psi by a factor of about 6 billion to 1, which is noteworthy even for a skeptical reader."[17]

In a different multi-university study published in 2014, researchers analyzed seven independent studies showing that the human body can apparently detect random stimuli one to ten seconds before the images or sounds are actually presented to the person. For example, when people were randomly shown pleasant and unpleasant photos, their bodies reacted to the unpleasant images *before* the picture was shown to them. This ability to "feel the future," even when the person was not consciously aware of it, appears to be a form of precognition that we can actually measure in a neuroimaging lab.[18]

Even though we don't seem to be conscious of this precognitive capacity, it may help us to deal with emergency and crisis situations. Evidence suggests that everyday consciousness (Level 3 in our Spectrum model) is rather slow, especially compared to other systems in our brain. Furthermore, the intuitive processes in our brain are much faster than our conscious ability to rationally think.[19] The areas that generate intuition are located in the insula and anterior cingulate cortex, and they contribute to our ability to understand the world in more global and comprehensive ways.[20]

Intuitive awareness (Level 5) is not language oriented, but it helps to inform our frontal lobe consciousness when it comes to moral and ethical issues.[21] This intuitive awareness is also part of the scientific process, where seemingly spontaneous realizations have given researchers the tools to understand the nature of DNA, gravity, geometry, and relativity. As Einstein once said, "The intellect has little to do on the road to discovery. There comes a leap in consciousness, call it intuition or what you will, and the solution comes to you and you do not know how or why. All great discoveries are made in this way."[22]

We suggest that Enlightenment shows a person that there is more to life than our intellect can see, and by interrupting our normal consciousness—reflecting on the creative levels of awareness that exist inside us—we can change our entire view of reality. This is where transformation occurs, whether it is a tiny insight or an "aha" experience, or a grand shift of awareness brought about through the experience of Enlightenment.

TONY ROBBINS AND THE ONENESS BLESSING

Personally, I am fascinated by the possibility that consciousness could affect something at a distance, and it would certainly add a new quality to our description of Enlightenment. Several years ago, I had the opportunity to do an experiment involving this concept with Tony Robbins, the well-known motivational speaker. It was exciting to meet him, his wife, and his small dog, and I had a lot of fun getting to know them.

Tony had become interested in a practice called the Oneness Blessing, a derivation of a Hindu spiritual tradition known as shaktipat, where wisdom and energy are directly transmitted from an enlightened teacher into the consciousness of a student. Once enlightened, you are then empowered to bestow the blessing on others.

Tony offered to give me and a few others the Oneness Blessing. I went up to the Four Seasons Hotel in New York City where a suite had been set up for about fifteen people to receive the blessing. Tony instructed us to find a comfortable position to sit in and to keep our eyes closed throughout the ceremony. It began with soft, meditative-like music, and we were asked to concentrate on our breath, paying attention as we slowly inhaled and exhaled, relaxing our bodies and minds. After about fifteen minutes, several people who were considered "Givers" walked up to us and placed their hands over our heart and on our head. You could feel their hands vibrating as they "transmitted" the blessing to us, but not a word was uttered. This was done about three or four times during the ceremony, each time lasting about sixty seconds. Then we were told to slowly open our eyes and bring our consciousness back to the present moment.

I actually had a surprising experience. Although my eyes were closed the entire time, when I was touched for the first time, a strong light came in through my eyelids. Imagine being in a darkened room and somebody suddenly shines a flashlight in your face while you keep your eyes shut. That's exactly what I experienced. It made me think of the many descriptions of Enlightenment in which people perceived a powerfully bright and beautiful light. While I did not have that type of intense experience, there were similarities. Now it's possible that the sun had simply come out from behind a cloud, making the room brighter, but with my eyes closed I couldn't tell if I imagined it or if it was related to the Oneness Blessing. I've never had that kind of visual experience again, so it's difficult for me to judge it, but I certainly was intrigued enough to set up a neuroimaging experiment.

I wanted to see what was going on in the brains of both people involved in the Oneness Blessing—the sender and the receiver. Nothing like this had ever been done before, and I was very excited to see what the scans would show. Would they be similar to our scans with the Buddhists and nuns, or would the brain look more like the mediums and Pentecostals who entered into trance states? Or would there be some other pattern, perhaps one more specifically related to enlightenment?

But I faced a problem: How do you set up a control situation where the receiver can't tell if a blessing is being given or not?

Obviously, if someone puts a vibrating hand on you, you'd know they were doing something. If I could figure out a way to "blind" the study, where the receiver couldn't tell when the energy transmission was taking place, I could see if there was different neurological activity associated with the Oneness Blessing. Then we could say that "something" had been transmitted from one person to another.

An interesting possibility came to light. Several of the practitioners informed me that the blessing could be given without actually being in physical contact with the other person. In fact, they said that I could put the receiver in a completely different room. That not only solved my problem, it provided a scientific opportunity to see whether a person's intention might influence the brain of another person at a distance.

I placed a small intravenous catheter in the arms of both a sender and receiver. Since they were in different rooms, all I had to do was inject the blood flow tracer into both of them at the moment when the Oneness Blessing was given. We made sure that the giver and receiver had no contact with each other prior to the study, but we gave the sender the person's picture, name, and location in the building. We told the receiver that we would conduct two sessions in random order—one would be the control or resting condition, and the other would be the actual transmission of the blessing. From the receiver's vantage point, there would be no way to know when the actual blessing was sent.

By using this SPECT protocol, as we did with the Brazilian

mediums, we could see the "before" and "after" pictures of the givers' and receivers' brains. If we found a difference, it would lend evidence to the possibility that some form of transmission or enlightenment took place.

THE POWER OF SENDING BLESSINGS

So what did we find? Normally, mentally concentrating on another person would increase activity in the frontal lobe. But sending the Oneness Blessing actually decreased frontal activity in the senders, a pattern very similar to what we observed in the Brazilian mediums and Pentecostals who spoke in tongues. The givers described their experience not as concentrating directly on their receivers, but as *surrendering* to the power behind the blessing, allowing that "energy" to flow through them to another individual.

Unlike the Pentecostals and mediums, we also saw decreases in the parietal lobe. Thus, their sense of self would dissolve as the divine "energy" of Enlightenment flowed through them. We saw similar parietal lobe decreases when people engaged in deep prayer for close to an hour. At that moment, the nuns felt as though they were in the presence of God or Jesus, and the Buddhists felt that they had merged with pure consciousness and their sense of self would disappear. Perhaps the same thing happens while giving the blessing: the practitioner's self disappears and only the spiritual energy—shaktipat—remains.

But we could not say whether or not the senders experienced

Enlightenment as they gave the blessing. Like the mediums, their belief system did not change, but perhaps it had when they initially *received* the blessing that turned them into people who now felt deeply connected with the Oneness Blessing.

THE POWERLESSNESS OF RECEIVING BLESSINGS

But what happened to the receivers? This, for me, was the bigger question, with far bigger stakes. Well, the results certainly were not impressive. When we asked the receivers which session they thought was the blessing, only 50 percent guessed correctly. Statistically, that's pure chance, no better than a flip of a coin. However, even if the people couldn't tell if they received energy from another person, I wondered if there might be value for the recipient.

When I looked at the brain scans, I found a few areas that were statistically different between the control condition and the Oneness Blessing condition. Activity in the right caudate and right hippocampus (areas involved with memory and abstract thought) decreased between 10 to 15 percent, so it's possible that these structures located in the deeper limbic areas of the brain might be affected by the Oneness Blessing. But if the recipient can't detect anything, how do we know if these subtle brain changes mean anything at all? Obviously the receiver didn't feel any of the qualities associated with enlightenment—intensity, clarity, unity, surrender, or a change in belief—nor did the receiver feel a sense of pleasure that would reflect changes of activity in the emotional

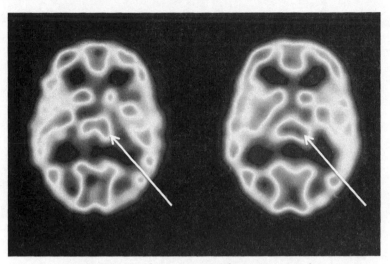

Figure 2. SPECT scan from a receiver showing increased activity
(arrow pointing to the darker region) in the thalamus.

areas of the brain, something a person would expect to experience if enlightenment was actually transferred to her.

Interestingly, activity in the thalamus increased in the receivers during the Oneness Blessing. As we have described before, this structure is involved with sensory perception, and it also relays information back and forth between the consciousness centers of the frontal lobe and the rest of the brain. So what was stimulating the thalamus? Perhaps there are subtle energies or sensations that our brain perceives but which never enter into our conscious mind. Given the fact that we only had a small group of people, and that the magnitude of the changes we recorded were very small, it is difficult to draw any firm conclusions from this experiment.

CHANGE THE CONSCIOUSNESS OF YOURSELF

Given the various qualities ascribed to an Enlightenment experience—feelings of unity, bliss, or transcendent insights—none of the recipients of the Oneness Blessing reported any unusual thoughts, images, or feelings. But perhaps it doesn't matter if consciousness can truly extend beyond the confines of the brain. What is more important, in my opinion, is whether the experience is perceived as positive or negative, beneficial or disruptive. What we've found, throughout all of our studies, is that most people feel that their spiritual practices add great meaning and purpose to their life. Without a sense of meaning and purpose in life, we are far more prone to anxiety and depression.[23] This is especially true for adolescents, for as psychologists at the University of South Florida found, a sense of purpose enhances "coping, generosity, optimism, humility, mature identity status, and more global personality integration."[24] With the rise of the "nones"—a term used to identify adults and the large percentage of youth who have turned away from mainstream religion and spirituality—we need to find new tools and experiences to enlighten the minds of the next generation of seekers.

We may not be able to instantly enlighten others with our hands or silent thoughts, but we can build belief systems that are filled with optimism, curiosity, open-mindedness, and desire to transcend the veils of ignorance. Practices such as meditation, mindfulness, yoga, or deep contemplative prayer, as the vast majority of the research has shown, help us to better regulate our emotions and increase our empathy and compassion for others.[25] The message is clear: focus on changing the consciousness of yourself and personal growth will occur. This might not fit the Eastern concepts of Enlightenment but it clearly reflects the Western notion exemplified by the Age of Enlightenment, where intense self-reflection would lead to the discovery of rational and scientific truth.

One final thought on the Oneness Blessing. We know that face-to-face interactions have a more powerful effect on the brain. Perhaps the hand-to-body contact used by spiritual healers functions in a similar way. Of the hundreds of studies conducted on healing touch and various "energy therapies," as they are referred to in the complementary medicine field, many have found decreases in a patient's anxiety, stress, and pain.[26] But again, the research is ambiguous, with many other studies finding no statistical significance. The reason, however, may be simple. The most beneficial changes are associated with practices where the patient becomes an active participant in her own healing ritual. The solution, then, is equally simple: create your own spiritual practice combining any technique you find pleasant, stimulating, and meaningful. And if you bring your personal, relational, and spiritual *values* into the words you speak and the actions you take with others, that, in my opinion, is the profile of an Enlightened individual.

EIGHT

Opening the Heart to Unity

t was eight p.m., and a small breeze filtered through the Big Sur mountains situated on the Northern California coast. A group of about twenty-five people were about to engage in a little-known Sufi ritual called Dhikr (sometimes spelled Zhikr and pronounced "zicker" with a soft "z"). Sufism is a mystical branch of Islam that focuses on divine love, unity, and enlightenment.

The class divided into three groups. The first group sat with their backs against the wall of the room. Their task was to rhythmically clap their hands or tap on a variety of drums while the second group formed a large circle. They would chant *lā 'ilāha 'illā-llāh*, which roughly translates as, "There is no god but God." It's an invocation of the oneness of everything and an invitation to open one's heart to the direct experience of God's love. While chanting, they would also rock their heads from side to side.

The third group gathered in the center of the circle to practice the dervish ritual of whirling, a ceremony created in 1273 by the Mevlevi Sufis located in Turkey. Those who would be whirling stood with one arm stretched upward and the other toward the ground, and they would begin to slowly turn in a circle as the outer group began to chant, clap, and drum.

As the speed of the drums increased, the chanters raised their voices and deepened the rocking of their head and torso. Those in the center who were whirling went faster and faster. Such movements have powerful effects on the brain,[1] and the rhythmic elements stimulate the autonomic nervous system, activating reward areas of the brain associated with strong positive emotions.

Kevin was one of the participants in the outer circle who were chanting and rocking. He was a forty-eight-year-old man who wavered between agnosticism and atheism, but he loved to explore the rituals of different religious groups. He enjoyed the experience, but then something unusual happened. Instead of hearing his voice coming from his mouth, it sounded as if it came from a place about three feet away, to his left. It intrigued him, but it also scared him because it felt like he was having a psychedelic experience and losing control over his senses. In Dhikr, the goal is to fully surrender yourself to the divine presence of God, not unlike what the mediums did when channeling the dead, or what the Pentecostals do when they invite the Holy Spirit to speak through them. Instead of going deeper into the altered state of consciousness, Kevin chose to sit down with the drummers. He felt as though his mind was floating in the clouds.

When the ritual came to an end, he found himself in a state of bliss, with pleasurable sensations racing through his entire body.

Later that night, he awoke from a dream in which he saw himself in a beautiful mosque illuminated by intense patterns of neon lights. But even though he had awakened and opened his eyes, the vision continued for another thirty minutes. As he sat in bed observing the colored lights, he heard a strange melody in the back of his mind. Suddenly he felt filled with a rare feeling of self-love that continued for over an hour. Then he fell into a very deep sleep, waking up totally relaxed and invigorated. The feelings lasted for several days, and as he described his experience to me, he said, "I felt like all of my worries and doubts just fell away! I've never felt such a sense of inner peace." Since Kevin was familiar with Eastern descriptions of Enlightenment, I asked him if the experience had changed his life in a fundamental way. Yes, the experience was intense, and yes, the sense of inner peace and self-love was particularly profound, but he soon found himself immersed in his old and familiar feelings of self-doubt.

"No, I don't think I was Enlightened," he told me, "but it showed me that I could reach such a state, and I've used that experience as a beacon to guide me to where I truly want to be."

I would say that this was a little "e" experience—a taste of what Enlightenment could be. For some, the same experience could change their entire lives, and for others—especially those more inclined toward skepticism—it might mean nothing more than an enjoyable and uplifting experience. That's what makes Enlightenment so fascinating to study. It can be sudden or gradual, and it depends on how valuable and meaningful the experience is.

Kevin's experience made me wonder what kind of brain activity Islamic and Sufi practices might evoke. I suspected that the brain would look similar to the pattern associated with other practices that lead to an Enlightenment experience, and I would soon have the opportunity to find out. But first, a little background about Islam, as it may be one of the most misunderstood religions on the planet, especially to nonfollowers.

TRADITIONAL ISLAM

In today's world, much of the confusion surrounding this fifteen-hundred-year-old religion is based on what we hear about extremists. As is true with all religions, many adherents are peaceful, but there are always certain groups who will use selected scriptural passages to incite violence toward others. When this occurs, it's easy to overlook the fact that most religions espouse very positive ideals such as the Golden Rule, the Ten Commandments, or the peaceful practices of meditation.

The term "Islam" comes from the Arabic verb *aslama*, which means "to accept, surrender, or submit," and the basis of Muslim spirituality is to open one's heart to the oneness of Allah, the Arabic word for God. As you can see, the notion of oneness and surrendering are two of the five main elements found in most Enlightenment experiences. However, many people incorrectly assume that God and Allah are different entities, but the Quran is modeled more closely to the Hebrew bible than to the New Testament.

GOD'S ORIGINAL NAME

Although one does not have to believe in God to reach Enlightenment, God often can provide an important focus for facilitating it. But it depends on how you define God. Think about it for a moment: What does God mean to you? Our research shows that very few people—less than 10 percent—define this mysterious word in a similar way!

The word "god" first appeared in a sixth-century Christian book called the *Codex Argenteus*.[2] Prior to that time, "theos" was used throughout most of the early Christian writings. It was a term used for any deity, and originally came from the Greek word "Zeus." The use of the word "allah" can be traced back to the eighth century or earlier, and continues to be widely used by Middle Eastern Christians, Jews, and Muslims. Most scholars agree that Allah is a variation of the Aramaic Elah and the Hebrew Eloah, or Elohim, a generic word that referred to the highest and most powerful of all the deities, two of whom—named El and Al—were written about in ancient Assyrian, Phoenician, and Babylonian texts.[3]

God and Allah are interchangeable in the sense that both refer to a singular deity who created and oversaw humankind. But since the true name of this deity could not be spoken outside the temple that was destroyed nearly two thousand years ago, indirect names were given. For example, God's name is sometimes referred to using the Hebrew letters "yod," "heh," "vav," "heh" (YHVH), an unpronounceable word.

In both Jewish and Muslim mystical traditions, there were

many names that were used for God: Infinite Knowledge, Perfect Goodness, All Powerful, Righteous, Supreme Being, Miracle Maker, Emancipator, Defender, All Wise, All Forgiving, Life-giver, and the Bringer of Death. In some contemporary mystical circles, Allah has been called the Source, the Breath, or the Oneness of everything. These mystical traditions recognized the indescribability of God as well as the importance of an Enlightenment experience in which people feel as if they "touched" God in some profound way. In the mystical paths of Christianity, Judaism, and Islam, the follower strives to feel a deep unification with God by surrendering one's self to this holy presence. When this occurs, it often leads a person into a powerful new belief system.

In Islam, the classic creed of the Shahada states that "There is no God but God." In Judaism, there is a foundational prayer, the Shema, which similarly states, "The Lord is God, the Lord is One." In the Christian faith, the Apostle's Creed reads, "There is only one true God, eternal, infinite, and unchangeable, incomprehensible, almighty, and ineffable." Thus, in the three Abrahamic traditions of the West, there is a mutual consensus in the oneness of God, and in that oneness there are hundreds of qualities and powers associated with this mysterious word.[4]

MYSTICAL ISLAM

Sufism can be traced back to Muhammad's cousin and son-in-law, Ali ibn Abi Talib. The early Sufis would delve into the Quran reciting key passages while remaining in a deep meditative state until they experienced a direct connection to the divine, the rev-elation of truth, and the ecstatic experience of pure love and peace, qualities very similar to the Hindu mystical practices of becoming one with pure consciousness.

As different Sufi brotherhoods were formed, each one adopted different rituals and theologies. Today you can find hundreds of small neighborhood groups throughout the Middle East, spreading from North Africa to Pakistan and India. Music, dancing, and the repetitive chanting of religious phrases are used to enter trance states, but many mainstream clerics consider such practices heretical, especially since Sufi leaders would often offer radical interpretations of the sacred texts based on their mystical experiences.

In the mid-twentieth century, Sufism became widely known through the English translations of Rumi, a thirteenth-century Sufi mystic whose poetry often reflected Eastern concepts of Enlightenment. Different styles of Sufism were introduced to Europe and America, and some even excluded much of the Islamic philosophy from which it emerged. Thus, Western Sufism became a devotional practice based on the concept of enlightenment through universal love, acceptance, and becoming one with the divine names of God. As is true with most mystical traditions, there is no central orthodoxy or unified doctrine.

For these reasons, many scholars do not consider Sufism a religion. Instead, they categorize it with other esoteric wisdom traditions (Native American spirituality, Celtic rituals, etc.) found throughout the world. For example, in early Christian gnostic practices, a person would be "enlightened by the divine Word."[5]

Sufis are not literalists, they are mystics, and strive for an Enlightenment which connects them deeply to God.

INSIDE THE SUFI BRAIN

Sufis employ a variety of techniques, like Dhikr, to achieve mystical union with the divine. The purpose of Dhikr is to remember and embrace the spirit of God, and it involves a series of complex rituals like the one I described at the beginning of this chapter. The explicit aim of Dhikr is to enter an ecstatic state (*hal*) that will purify the heart and open the practitioner to spiritual intuition (*qalb*). This helps the person overcome the demons—real or imaginary—that fill the worldly mind (*nafs*). When, through practice, a person can extinguish the individual personality and ego (*fana*), one gains deep spiritual knowledge that comes from direct communion with Allah or God (*haqiqat* and *marifat*).

In another Dhikr ritual, the aim is to reach Enlightenment by moving through the stages of the heart as you experience the ninety-nine names of God. As the eleventh-century Islamic theologian Al-Ghazali reflected in his book *The Marvels of the Heart*, "When God becomes the ruler of the heart . . . there shines in it the real nature of divine things."[6] Connecting with God results in an awakening—an Enlightenment—in which the person comprehends the true nature of the world. Many historians view Al-Ghazali as the second most influential Muslim after Muhammad, laying a foundation for the general acceptance of Sufism within Islam.[7]

Many books have been written on Sufism, but to my knowledge no brain-scan studies have ever been done on their spiritual practices. With the help of several of the local Muslim associations connected with my university, I found a group of Sufis and Muslims who were willing to perform their rituals in our lab.

Since Dhikr is usually performed in a group, I brought two practicing American Sufis into a hospital room where they could do their practice together using a variety of incantations and body movements. They chanted and swayed for an hour, similar to the ritual Kevin participated in, and then, using our SPECT protocol, we injected them with our radioactive tracer during the end of the practice, when the individuals were presumed to be in the deepest trance-like state most likely associated with enlightenment experiences. We then took pictures of the changes in their brain.

The Sufis showed dramatic decreases of activity in the frontal lobe, similar to what we found with the Pentecostals, who, by the way, also engaged in intense body movement. As we have described, when you decrease activity in the frontal lobe, it becomes easier to access creative states of imagination—Level 4 on the Spectrum of Awareness. Visions may occur, voices can be heard, and out-of-body sensations may be felt—similar to what Kevin experienced—and they can feel intensely and extraordinarily real, lasting for many hours or days.

But is this enough to achieve Enlightenment? Perhaps, but we believe that a person must also deeply reflect (Level 5) on the experience to assess the meaning it has for one's life.

I also made another fascinating discovery. In the Sufi scans, we saw greater decreases in the activity in the right frontal lobe when compared to the left side. We saw the same thing with our mediums, and other researchers have noted similar brain asymmetries in their meditation studies.[8] The right frontal lobe is frequently involved with negative thinking and worry. A decrease signifies less pessimism and thus an improvement in emotional health.[9] This explains why most people experience positive feelings like joy and bliss when they meditate or engage in intensive practices like the ones we've described in this book.

Consider the significance of this for a moment. From a neuroscientific perspective, intense spiritual practices actually change our ability to perceive the world around us. Areas of the brain that are normally dormant when we perform our daily tasks can come online during ritual practices. Our sense of reality changes and this allows the brain to form new neural connections. Old habits can be suddenly interrupted, allowing us to form healthier behaviors. This gives us greater freedom to change our outlook on life.

Gentler forms of meditation and prayer lower stress and anxiety, but they rarely have the power to dramatically alter our consciousness in the way that intense practices like Sufi chanting can do. Instead, contemplative practices like mindfulness allow us to *accept* ourselves for who we are. But if we add more intense practices involving deeper breathing with repetitive movements and sounds, we can break down the neurological circuits that keep old beliefs firmly rooted in place and transform the ways we think and behave.

An Enlightenment experience radically rearranges many neuronal connections in a relatively short time. The result is a tremendous benefit to our brain and body as we discover new positive ways of thinking, feeling, and experiencing the world around us.

In addition to the frontal lobe findings on our brain scans of Dhikr, we found that parietal activity declined after fifty to sixty minutes of chanting, similar to what happened when our Christian and Buddhist meditators engaged in hour-long practices. Since this area is responsible for constructing and maintaining our self-image, a sudden decrease helps to explain why many spiritual practitioners say that their sense of self—their ego—seems to disappear, an important element often talked about in Eastern philosophy. All that remains is the object of your contemplation: the unity with Allah, a merging with the universe or pure consciousness, or feeling closer to God. As the Islamic texts proclaim, in that moment "there is no God but God." From a neuroscientific point of view, you might say, "there is no awareness but awareness," or "there is no love but love," perspectives echoed in the meditation traditions of Buddhism, Hinduism, and Christianity. It is a neurologically "real" state of being-ness that promotes inner and outer peace. Unity consciousness allows a person to feel intimately connected with everyone and everything, and the love that bursts forth from one's heart is one of the essential goals of the Sufis.

PRAYING OUT OF HABIT

Dhikr is associated mainly with Sufism, but in the larger Islamic community there is another form of prayer called Salat, which Muslims are required to do several times a day. Salat is a repetitive ritual involving standing, bowing, sitting, arm and hand movements, and prostration while the person recites various verses and prayers. It takes between ten to twenty minutes to do, and it is considered a sin to pray less than five times a day.

When it comes to the brain studies of Muslim spiritual practices, the research is sparse. One study, conducted at a Malaysian university, found no significant alterations in alpha brainwave activity (associated with relaxation) as Muslims performed the four required cycles of movement.[10] This suggests that habitual prayers, in any religious tradition, may have little effect on the brain. However, the researchers used EEG devices that can only measure weak neuroelectric activity on the outermost surface of the brain. There may have been significant changes taking place in deeper structures. That's why we prefer to use SPECT and fMRI scanners (which construct a moving picture of neurological activity), even though such experiments can be expensive to conduct.

Several studies have found benefits from doing a daily practice even if it is habitual. For example, a recent study showed that practicing Muslims who performed daily Salat had superior dynamic balance (an important physiological function, especially for elders) over nonpracticing Muslims.[11] In another study conducted at an Iranian university, forty-five minutes of Islamic prayer significantly reduced anxiety for Muslims who were about to go into surgery.[12]

I recently had the opportunity to perform a brain-scan study on Salat, and I wanted to see if there was a difference between doing the prayer in a very "automatic" manner (without much actual effort) compared to doing it more intentionally, with greater passion, intensity, and concentration. My first subject—I'll call him Ibraham—was not a highly religious person, although he was a practicing Muslim. He felt, as do we, that deeper practices are more likely to result in the feeling of surrender that could lead to Enlightenment.

ARE THERE ENLIGHTENMENT "BUTTONS" IN THE BRAIN?

Few studies have looked at what happens when a particular type of prayer or meditation is practiced at varying levels of intensity, as we did with Ibrahim. I did something similar a few years ago when I studied highly experienced meditators performing two related types of Kundalini yoga practices.[13] One was called Kirtan Kriya and the other was called Shabad Kriya. They both involve the repetition of the mantra *SA-TA-NA-MA*, but the Shabad form includes very deep breathing. According to Yogi Bhajan, who introduced these techniques to Americans in the 1970s, Shabad Kriya could awaken the spiritual energies in the body that result in Enlightenment. The technique is very powerful, and some people have even experienced temporary symptoms resembling psychosis.[14]

In our study, we initially saw increases in the frontal regions, but they were followed by significant decreases as the intensity of the meditation practice increased. This is consistent with our theory of what occurs in the brain when a person experiences a profound spiritual or mystical state. We also observed decreased activity in the parietal lobe associated with an increasing sense of oneness and connectedness. Activity in the limbic system also increased, explaining why our subjects experienced more intense emotions. This is similar to what we saw in the Sufi practice of Dhikr. I wondered if we might see a similar change with the daily ritual of Salat. When we scanned Ibraham's brain during the routine practice, we saw some general activity in the frontal lobe, but nothing particularly active. In other words, the brain was responding in the way it would to any habitual routine, like combing your hair or driving a car. When an action is memorized and repeated, the frontal lobe doesn't need to be very active.

Then I asked Ibraham to deliberately increase his concentration as he performed the prayer again, generating as much intensity as he could. I wanted him to consciously try to seek a revelatory experience, similar to those described in the Quran. So he performed his practice as intensely as he could. He slowed the process down, focusing on every word and every movement. He concentrated harder than he ever had before to make this prayer more powerful. Afterward, he was visibly spent, and when I asked him how it felt compared to his usual way of doing Salat, he said that he felt great joy and a deep sense of surrendering to God.

When we scanned him during this more intense practice, there was a substantial decrease in frontal lobe activity, similar to the trance practices in which the person feels a sense of surrender. In addition, we noticed an increase in the anterior cingulate cortex (see Figure 3). This was interesting because these two brain areas tend to go together, so why would they go in different directions?

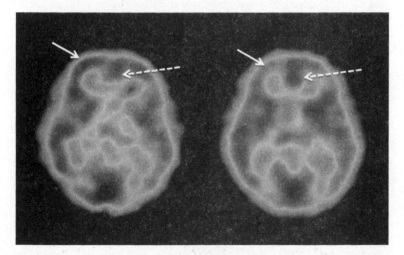

Figure 3. The image on the left shows the activity of the brain during the habitual practice of Salat. As Ibraham intensified his practice, we see, in the scan on the right, decreased activity in the frontal lobe (the solid arrow shows that the area that was initially dark is now much lighter—less activity—during the prayer). There is also increased activity in the anterior cingulate (the area indicated by the dotted arrow on the right shows up darker on the scan).

Normally, the anterior cingulate cortex helps to regulate emotional responses by receiving signals from the frontal lobe. It calms you by turning down activity in the emotional centers of the brain. Instead of feeling calmness, which is what we see when both the frontal lobe and anterior cingulate cortex are stimulated, Ibraham experienced powerful feelings of emotional intensity and surrender. Since intellectual worries and fears could also be interrupted by frontal lobe decreases, this would explain why intense spiritual practices are more associated with feelings of bliss and ecstasy. There were also increases in the brain's reward system areas consistent with positive emotions. When these areas are stimulated, dopamine is released, a powerful pleasure neurochemical.[15] We also saw decreases in activity in the parietal lobe, which did not occur during habitual prayer. Thus, having a clear conscious *intention* to immerse oneself in a spiritual practice can increase feelings of unity and connection.

DON'T MEDITATE ON YOUR DISBELIEFS

Mark, my coauthor, who has experimented with many spiritual practices, had a similar experience when he participated in one of our earlier experiments. He was focusing on an image of God—a radiating person with long white hair. Suddenly the image disappeared, and feeling somewhat disconcerted, Mark deepened his breathing and envisioned a white light pouring through him, an image used in many Eastern practices. We had previously arranged for Mark to signal me when he felt that he reached a deep contemplative state. Normally, that takes about forty to sixty minutes, but he raised his finger after about six minutes. When I told him how short it had been, he was genuinely surprised, saying that it felt like an hour. He described the experience of being filled with intense energy, and we know from other researchers that time distortion is common during hypnoses, ecstatic states, flow experiences, and experimenting with various drugs.[16] Furthermore, abrupt changes of consciousness are associated with changes in many parts of the brain.[17] On the scan, we saw a substantial increase in one half of Mark's frontal lobe, and a substantial decrease in the other half. Very strange! In fact, for three days, Mark felt so high he could barely sleep. At the time, he believed that it was one of the most intensely positive experiences he had ever felt—a genuine Enlightenment experience.

Then his mood suddenly changed and he felt exhausted for the next couple of weeks. Mark's interpretation: in his desire to alter his consciousness, he triggered a state of intense excitement, similar to a manic episode. He returned to "normal" and didn't feel that the experience enlightened him in any way.

But in some ways I would disagree, because Mark, over the next few months, changed his entire strategy concerning his practice of meditation. Instead of using traditional practices that contained spiritual imagery and theologies he didn't really believe in, he chose instead to focus on key secular values that gave his life more meaning and purpose, concepts like compassion, peace, and self-acceptance. Not only did his behavior change in a way that improved his life, he realized that meditating on disbeliefs could cause abnormal activity in the brain. His insights dramatically changed the direction of his life and his research, and that fits all of the criteria of Enlightenment.

You see, Mark had never really believed in the existence of God, but he was fascinated by the concept. His parents were Jewish, but they rarely attended temple, nor were religious perspectives brought up in family discussions. I suspect that intense focusing on an object or belief that is not core to your value system can cause neural dissonance, an unpleasant state where emotional conflicts can arise. His brain scans suggested that a person can consciously hold conflicting values and beliefs, but the conflict doesn't allow the brain to integrate the experience in a meaningful or productive way.

I shared my opinion with Mark, and although he agreed that all of the criteria for Enlightenment seemed to have been met, he didn't want to call it that. I fully understood. To claim that one is Enlightened feels narcissistic. Mark did not feel that his personality had changed, only that his orientation toward meditation had been altered. It certainly was a big "aha" moment in his life, so perhaps one could call it a little "e" experience.

Mark's experience reminds me of what happens in Zen practice when the student is given an incomprehensible puzzle—a koan—to solve. There is no way of using your logic or reason to respond to the Zen master, and if you try to answer in a "normal" way (using the everyday consciousness of Level 3), the teacher might whack you with a stick. So no matter how hard you try to focus on the problem, no solution can be found. When the brain is faced with a seemingly unsolvable situation, we experience neurological distress, but some of the newest research shows that it also promotes sudden insight. As brain researchers at Drexel University found: "Insight occurs when a person suddenly reinterprets a stimulus, situation, or event to produce a nonobvious, nondominant interpretation."[18]

In my opinion, this is exactly what happened to Mark. He (or his brain) couldn't tolerate the discrepancy between what he thought he was doing and what the scans actually showed. The neural dissonance caused Mark to pursue a different and more personally fulfilling spiritual path, one that wasn't burdened with his biblical images of God.

Based on our brain scan studies of other atheists, I have come to the conclusion that it is generally unwise to focus too much on disbeliefs because it can generate anxiety. My advice: when choosing to pray or meditate, focus on those desires, beliefs, and values that mean the most to you. If you are an atheist, don't pray to God but meditate on a beautiful rainbow, or love, or pure consciousness, or the simple joy of being alive. If you are religious, consider praying to God more intensely than you have before. If you are a parent, don't ruminate on your child's weaknesses; meditate on your child's strengths. If you are political, don't focus your distaste for the other party. Instead, demonstrate with your whole being the application of those values that are sacred to you.

But don't forget to try something new. Experiment, like Mark did, with other religious values. Try to see the world through different eyes. When you do so, you become more tolerant of others and more loving of yourself, because openness increases your likelihood of experiencing Enlightenment. You may just experience the world in a totally new way.

STRUGGLING WITH THE DARK SIDE

But what about the potentially negative side of spirituality and religion, as is so often cited by atheists and attested to by acts of terrorism by religious groups around the world? Can Enlightenment itself be associated with negative emotions such as anger, fear, or violence? There are examples in which Enlightened individuals engage in violence. Most famously, the Bhagavad Gita, an ancient Hindu text, describes a man named Arjuna, a prince who encounters Krishna, an Enlightened deity. They meet on the battlefield right before a bloody war is to begin. Arjuna refuses to go to war, unwilling to fight against his kindred, but in the end, he realizes that he must surrender to his warrior nature. It is viewed as an allegory reflecting all of the moral and spiritual dilemmas a person must struggle with throughout life in order to reach liberation and Enlightenment.

Battle metaphors exist in the Hebrew bible, the New Testament (Revelation), and the Quran (jihad), and there are always two opposing interpretations: one literal and the other metaphoric. All mystical traditions, including Sufism, invoke the latter by stating that everyone must do battle with the inner demons of their mind. To quell the turmoil we are born with—to end the cosmic war of the soul—that is true Enlightenment.

Recently, many neuroscientific studies have explored how morality works in the brain.[19] Not surprisingly, we find that the areas stimulated by intense prayer and contemplative meditation are the same ones that help us distinguish between good and evil, generosity and greed, and love and hate. We believe that spiritual practices neurologically force us to confront these issues in our lives, and as many spiritual practitioners will tell you, one can spend years struggling with these emotional wars—the inner jihad, as the Sufis say—in order to reach inner and outer peace.

The story of Arjuna reminds us that Enlightenment does not mean an end to violence and conflict, but rather a new understanding of the nature of conflict in the context of a greater reality. Knowledge of how our brain processes our experiences—from despair to Enlightenment, from separation to unity—may help to bridge the gap between science and religion. And if we use our brain science wisely, we can take advantage of spiritual practices like Dhikr, the Oneness Blessing, and even mediumship to enhance our love and compassion for others. That is the path of the Sufi, a path that opens the heart and mind:

I ruminate on God

And my old self falls away.

Am I a Christian, a Hindu, a Muslim, a Buddhist, or a Jew?

I do not know for Truth has set fire to these words.

Now they are nothing but ashes.

I ruminate on God

And my old self falls away.

Am I a man, a woman, an angel, or even a pure soul?

I do not know for Love has melted these words away.

Now I am free of all these images

That haunted my busy mind.[20]

—HAFEZ, FOURTEENTH-CENTURY ISLAMIC POET

NINE

Believing in Transformation

Throughout history, spontaneous experiences of Enlightenment have happened to people in every culture. And as we have seen in the previous chapters, many different types of spiritual practices can provide paths toward higher states of consciousness (Levels 5 and 6 in our Spectrum). But our research suggests that people who actively seek Enlightenment are far more likely to experience it than those who don't believe in the concept. Actively pursuing transformational practices like compassion meditation or prayer or even just challenging your current beliefs in a deeply contemplative way, much the same way that I did on my way to Infinite Doubt, can all be useful approaches on the path toward Enlightenment.

The explicit intention to have a life-changing or life-enhancing experience primes the brain for Enlightenment, and the way you frame your intention will shape the meaning, quality, and intensity of what actually occurs. So I'd like you to take a moment to ask yourself these questions:

1. Do you *believe* that Enlightenment is possible?

2. Do you *desire* to have an Enlightenment experience?

3. Do you have a *strategy* to elicit such experiences?

Before you can intentionally pursue this goal, you'll need to try to define what Enlightenment means to you and clearly state what you hope to experience (peace, unity, clarity, connection with God, etc.). So take a moment right now, grab a pen and a sheet of paper, and answer the following question. Deliberately use your logic and reason, a process that causes you to think very slowly: "What does Enlightenment mean?" Take your time and write out a clear, concise, descriptive, and "accurate" definition.

Now try this experiment. Instead of rationally and logically *thinking* about the question, ask your *intuition*—a neurologically faster and free associative process—to answer this question: "What does Enlightenment *really* mean to me?" The less you try, and the more you observe the fleeting thoughts that flow in and out of consciousness, the more likely you'll come up with a surprisingly different answer.

Different people will answer this question in different ways. For example, one person may see enlightenment as freedom from anxiety or self-doubt. For another, it may symbolize a direct connection to God. But these are usually *intellectual* responses generated by the everyday decision-making processes of the brain (Level 3 in our Spectrum of Human Awareness). But when we ask our students and subjects to listen to their intuition, we are guiding them into Level 4, creative imagination. The same person will often "hear" or receive a different definition, like this: "I see enlightenment as a ball of light filling my consciousness," or "Enlightenment, for me, means to be free of my past." Then, by reflecting on both the logical *and* intuitive answers, you engage in a more comprehensive form of self-observation associated with Level 5 awareness, where the brain processes information visually and symbolically rather than with abstract concepts and words. Sometimes this exercise alone can lead to a small "e" enlightenment experience.

Also, the more clear you are about the change you would like to bring about, the more you prime your brain to search for that experience. So ask yourself this question: "What do I *really* want to discover that could change and enrich the direction of my life?" The more deeply relaxed you are, the easier it will be to hear those "whispers" of intuition. Write down your "enlightenment" or "transformation" goal and then meditate on the words, allowing your mind to wander and daydream. When a new piece of information comes into your awareness, write it down and continue to observe your thoughts and awareness. The longer the time you spend, the more insights you will come across, and as you continue to gaze at the words on your list, your brain will begin to seek experiences that will guide you toward your direction and goal. This, we believe, is the *conscious* way that will increase your likelihood of having a small "e" or big "E" experience.

PRACTICING SURRENDER

But the path for finding Enlightenment is never straightforward. If you genuinely want to experience a transformational shift (Level 6), you must be willing to temporarily suspend your beliefs about Enlightenment, including the ones you just wrote down. When you deliberately do this, your brain begins to search for a new belief to replace it. You are asking for Enlightenment, and if you continue to interrupt your old ideas, you push your consciousness into the higher levels of our Spectrum model.

You'll have to experiment because the process is unique for each individual, but the research shows that your brain can take action on any request you give it. For example, I know that in my mind, I was able to doubt everything I ever believed, and my brain would, to some degree, begin to distrust perceptions it normally takes for granted. You just have to trust your intuition, and you have to have faith. So spend some time questioning your deepest beliefs and focus your mind on what you don't or can't know about the universe or God. Give in to the uncertainty of not knowing and then "allow" the experience to unfold as you intensify any meditation you are familiar with (we'll guide you through several styles in Part 3).

I am often reminded of one of my atheist friends who once said to me that he would only believe in God if God came to him, shook his hand, and said, "I exist." I responded, "I'm willing to bet that you still wouldn't believe in God if that happened; instead you'd probably check yourself into a mental institution!"

You see, his disbelief in God was so strong that even if God punched him in the face, he probably would not believe that the experience was real. I see the same thing in scientists all the time, the staunch refusal to accept new ideas, or even evidence-based data that challenged their beliefs about the nature of the universe or how medicine should be practiced.

As a teacher, I always encourage people to think about alternatives, to stay open to new ideas, and to even play the devil's advocate against one's own beliefs because it's very good for your brain. Challenging your beliefs and opening your mind to new ones can be a very intense experience, but don't judge it. Just let the process unfold in its own way.

Evidence suggests that no matter what you think Enlightenment might be, the actual experience is usually very different from anything you could imagine. At some level you must be willing to accept whatever the experience brings. In most religions, this is referred to as surrender, or giving your will over to some higher authority or power. This requires faith, perseverance, and devotion.

But giving up old beliefs involves risk. So religion poses a double bind: traditions demand that you adhere to the specific tenets of the organization, but Enlightenment involves transcending them. This partly explains why new religions typically are established by people who felt enlightened by their spiritual endeavors, and it also explains why the orthodoxy will persecute them. And when your beliefs are transformed, it appears to be neurologically impossible to return to the old ones. This is what can happen to people who have powerful experiences that suddenly shift them from religiosity or agnosticism to atheism, and when that occurs, they often cannot return to practices focusing on God. After all, if you just went back to your old beliefs and habits, it wouldn't be much of a transformation.

TOLERANCE AND ACCEPTANCE

This raises a deeper question: *How* does Enlightenment change your beliefs? As we described earlier, our survey participants described many permanent changes in their lifestyles and attitudes. For example, 90 percent felt more spiritual, but only 50 percent felt more religious. This suggests that Enlightenment forces you to transcend your traditional religious doctrines. Seventy-five percent also reported a reduced fear of death, and 80 percent said they felt that their lives had more purpose and meaning. But perhaps the most important change I found was that Enlightenment seemed to help people become more accepting of others. So it stands to reason that those who are more accepting of different beliefs will have a better chance of actually finding Enlightenment.

In order to test this more directly, I developed a questionnaire with my colleague Nancy Wintering, called the "Belief Acceptance Scale," designed specifically to measure the degree of openness a person had toward different religious and cultural belief systems. We included it as part of our larger survey so that we could understand how openness related to different types of experiences. You can take it right now. Simply answer these nine questions as truthfully as you can, and circle the number of your answer. When you are done, add up the total of the numbers you circled. For example, in question 1, if you strongly believe that all religions are fundamentally correct even though their belief systems may differ from your own, you'd circle answer 4, giving you 4 points.

However, question 2 is a little trickier because it asks you to select *all* of the different ceremonies you participated in. If you circled only one of them, you'd get only 1 point, but if you circled all five, you'd give yourself 5 points (the maximum score you can have for question 2).

BELIEF ACCEPTANCE SCALE

Instructions: The following questions concern your spiritual or religious beliefs and experiences. There are no correct or incorrect answers. For each question, select the number of the answer that is most true for you.

1. I THINK OTHER RELIGIONS ARE CORRECT EVEN THOUGH THEY DIFFER FROM MINE.

   1. definitely disagree
   2. tend to disagree
   3. tend to agree
   4. definitely agree

2 . PLEASE SELECT ALL CEREMONIES YOU HAVE ATTENDED
OUTSIDE YOUR RELIGIOUS OR SPIRITUAL GROUP.
1. initiation rituals for infants, children, or adolescents
2. weddings
3. healing rituals
4. ordination into a religious vocation
5. funerals

3 . HAVE YOU ATTENDED WORSHIP SERVICES OUTSIDE YOUR
RELIGIOUS OR SPIRITUAL GROUP?
1. never
2. once or twice
3. several times
4. often

4 . HAVE YOU ACTIVELY PARTICIPATED IN WORSHIP SERVICES
OUTSIDE YOUR RELIGIOUS OR SPIRITUAL GROUP?
1. never
2. once or twice
3. several times
4. often

5 . HOW COMFORTABLE WERE YOU WITH THE IDEOLOGICAL/
THEOLOGICAL CONTENT OF WORSHIP SERVICES OUTSIDE
YOUR RELIGIOUS OR SPIRITUAL GROUP?

1. not at all

2. somewhat

3. moderately so

4. very much so

6. HOW COMFORTABLE WERE YOU WITH THE RITUALISTIC CONTENT OF WORSHIP SERVICES OUTSIDE YOUR RELIGIOUS OR SPIRITUAL GROUP?

1. not at all

2. somewhat

3. moderately so

4. very much so

7. IF YOU WERE DATING, WOULD YOU DATE SOMEONE OUTSIDE YOUR RELIGION OR SPIRITUAL BELIEF SYSTEM?

1. definitely disagree

2. tend to disagree

3. tend to agree

4. definitely agree

8. IF YOU WERE CONSIDERING MARRIAGE, WOULD YOU MARRY SOMEONE OUTSIDE YOUR RELIGION OR SPIRITUAL BELIEF SYSTEM?

1. definitely disagree

2. tend to disagree

3. tend to agree

4. definitely agree

9 . IF YOU WERE CONSIDERING MARRIAGE, WOULD YOU MARRY
SOMEONE WHO DOES NOT SHARE YOUR RACIAL OR ETHNIC
HERITAGE?

1. definitely disagree

2. tend to disagree

3. tend to agree

4. definitely agree

What was your total score? The maximum score you could achieve would be 37, which would mean that you are totally open to everything. The lowest score would be an 8, meaning that you are basically closed off to any other ideas and would have no interest in participating in ceremonies from different religious traditions.

When we looked at our survey respondents, the average score was 22. This is basically right in the middle of the highest and lowest possible score, so our survey suggests that people are generally ambivalent when it comes to accepting other people's religious beliefs. Those who have higher scores may be more likely to have big "E" experiences, those in the middle would be more inclined toward small "e" experiences, and those near the bottom would be least likely to experience any major shift in behavior or belief. Of course, all questionnaires are limited, so the results we found should be viewed as indications, not fact. The results showed that the more strongly positive a person felt about her own religion, the less tolerance she showed toward other religious beliefs. However, in countries like America where religious diversity keeps growing, we tend to be more accepting of others. Our study also found that those with higher levels of education showed more acceptance than those with lower levels of education. As people read and learn more about different traditions, and study with others from different backgrounds, the exposure to new ideas can broaden and change one's core beliefs.

Researchers at Harvard identified three qualities of beliefs, breaking them into facts, preferences, and ideology.[1] Try this little exercise. Take a moment to identify several of your strongest beliefs—both religious and nonreligious—and write them down. Think about the beliefs you have about work, family, relationships, politics, science, or morality. Which of them do you consider facts, and which do you accept as true because others said they were? And which of your beliefs do you simply *prefer* to embrace?

You might, for example, not believe in God. Or you might believe in God but not in the immaculate conception. And you probably believe that lying and stealing are wrong, no matter what your religious beliefs are. Those beliefs you consider to be absolutely true will establish the ideology that governs future actions and behavior. But those that are preferences are easier to question or suspend.

The Harvard psychologists discovered that children as young as seven can begin to view religious beliefs as preferences, not truths, and thus are more accepting of others professing different beliefs. In our survey, we also discovered that the older you get, the more accepting you become. This was especially true for those above fifty.

SUSPENDING OLD BELIEFS

Beliefs are essential for our lives, and yet we rarely call them into question. Nor are we aware of how they may limit our interactions with others. For example, we are all unconsciously biased toward thinking that our own beliefs are truer than others, and so we tend to judge other people's beliefs through our own egocentric attitudes.[2] In fact, a wide range of belief biases are associated with increased activity in the frontal and parietal regions.[3] But Enlightenment experiences appear to decrease activity in those areas of the brain, suggesting that the quest for transformation can, in fact, increase our ability to be more accepting of others.

Of all the questions answered in our survey, I found that most people had the highest degree of acceptance when it came to marrying someone from a different ethnic background. Most people expressed the lowest degree of acceptance when it came to participating in other religious practices. After all, it's difficult to listen to a sermon that you don't really believe in.

Our Belief Acceptance Scale also found that overall tolerance may be correlated with one's religious affiliation. People affiliated with Western monotheistic traditions had significantly lower scores of acceptance compared to those who were affiliated with Eastern traditions, where Enlightenment is the central theological belief. Individuals in Eastern traditions were also more willing to marry someone outside their belief system.

What about nonbelievers? Atheists were actually right in the middle of the pack, suggesting that some are very open to other people's religious beliefs, while others are adamantly opposed to any type of religious belief.

We did not find differences between low-, middle-, and upper-class individuals, so being poor or wealthy does not appear to have an impact on being open-minded, tolerant, and accepting of others. Gender also made little impact: men and women seem to be equally open-minded and closed-minded.

But when it comes to Enlightenment, those who reported the most intense experiences also had the highest scores on the Belief Acceptance Scale. Even more interesting was this finding: those who practiced meditation had higher acceptance scores than those who practiced prayer or engaged in other types of spiritual practice. So it seems that some practices, like meditation, help contribute to a sense of openness while others might not.

KINDNESS ENHANCES ACCEPTANCE

When people experience Enlightenment, they commonly report feeling more compassionate toward others. So developing kindness and compassion appears to promote the path toward attaining Enlightenment. In our previous studies, we have seen increases in acceptance when people practice compassion-based meditations. This led me to explore a specific practice called "Lovingkindness" meditation. The basic steps are simple. First, you send kindness to yourself by repeating, either out loud or in silence, any version of this phrase for approximately five minutes: "May I be happy, may I be well, may I be filled with love and peace." Next, you visualize friends and family members, saying, "May you be happy, may you be well, may you be filled with love and peace." Then you send this blessing to distant acquaintances, and then to people who have hurt and angered you. In the final step of this powerful exercise, you extend your love and kindness to everyone in the world—to all cultures, all colors, all religions, and all political groups. As you do this, you envision everyone getting along with one another and living together in peace.

Sending kind thoughts to difficult people who have caused you to suffer can be very hard to do, but it will change your brain in beneficial ways.[4] Your brain is designed to deeply embed negative memories, and each time you recall the insult or offense, you strengthen the negative associations of that memory as your brain sends out a distress signal to the entire body. In order to disable that association, you have to create positive thoughts and remain deeply relaxed when you visualize the person and the specific event that upset you. Eventually, the sensations of relaxation and the kind thoughts you generate will be consolidated into the old memory. But the memory will resist change, so you have to do this Lovingkindness practice often—sometimes for months or even years—before you'll experience calmness when you think about painful events from the past. Ultimately, those feelings of love, kindness, and forgiveness for the individual will become embedded into your memory circuits, enhancing feelings of compassion for both the person and yourself, and for everyone else you interact with. Again, our research suggests that gentle forms of meditation, mindfulness, and Lovingkindness practices help prepare the brain for Enlightenment.

Reflecting on forgiveness and compassion lowers your heart rate and blood pressure. Lovingkindness builds neurological empathy, acceptance, and compassion toward others.[5] Researchers at Yale and Michigan State documented that it reduces negative biases we unconsciously hold toward others,[6] and it increases the volume of key areas in your brain.[7] When I studied an individual who practiced Lovingkindness meditation every day for a month, I found a change in the way her brain functioned throughout the day. She activated her emotional centers and social centers far more robustly after doing the practice for a month, and she reported feeling much more open and compassionate toward others. I recommend that you practice this simple meditation for several weeks, because it will immediately make you feel more loving and accepting toward yourself.

Fostering feelings of love, forgiveness, and compassion can also lower your body's stress load by decreasing the cortisol.[8] Lowering your cortisol levels helps boost immune function and allows the brain to work more effectively.

FORGIVENESS TRANSFORMS THE SELF

Forgiveness is an implied element in the Lovingkindness meditation, but the conscious act of forgiving others can have a profound
effect on one's life and the brain.[9] In a series of studies conducted
at the University of Kansas, researchers found that for people who
practiced forgiveness toward those who offended them, their positive behavior spilled over into a wide range of social situations.[10]
They were more relational, and often used language like "we" and
"us," demonstrating a movement toward acceptance and social
unity. These people were more likely to donate to charity and vol-
unteer their time to others, qualities that psychologists refer to as
self-transcendence and that are often associated with the sense of
unity found in spiritual Enlightenment.

Forgiveness also provides many physical and psychological benefits. Fatigue decreases, sleep quality improves, and the quantities
of medication can sometimes be reduced.[11] You'll experience more
satisfaction with life, your mood will improve, and you'll feel a
deeper connection toward those you care for and love.[12] In fact,
a meta-analytic review of 175 studies involving 26,000 subjects
found that forgiveness was one of the most powerful tools for rein-
stating psychological balance, fortitude, and serenity.[13] For this
reason, we consider forgiveness as part of the "recipe" for inducing
Enlightenment because it can transform the personality of the
practitioner.

BECOMING A BETTER BELIEVER

Since our definition of Enlightenment suggests that a person may be limited by old beliefs, it would benefit us to consciously train ourselves to get past unconscious biases. So bring out your list of beliefs and consider these seven strategies that the CIA uses to teach its intelligence-gathering community to think more open-mindedly: [14]

1. Identify your own beliefs and recognize they are biased.

2. Become proficient in developing alternative points of view.

3. Do not assume that the other person will think or act like you.

4. Imagine that the belief you are currently holding is wrong, and then develop a scenario to explain how that could be true. This helps you to see the limitations of your own beliefs.

5. Try out the other person's beliefs by actually acting out the role.

6. Play devil's advocate by taking the minority point of view. This helps you see how alternative assumptions make the world look different.

7. Interact with people of different backgrounds and beliefs.

On the surface, these suggestions seem easy, but they're not. Take, for example, strategy 4 and apply it to the religious beliefs you outlined in the first exercise of this chapter. Can you even imagine that they could be wrong? Assume that they are wrong and ask yourself, "How could I have been mistaken?" If you believe in God, imagine how your behavior would change if you found out that God did not exist. If you're an atheist, and you discovered that God was real, how might that change your life? Often a person discovers that their basic values remain the same, and so they continue to behave as morally as they did before.

Apply this same step to your political beliefs and imagine, for a moment, that they could be entirely wrong. If you do this when you meet someone with a different political belief, you might be able to see that their reality is deeply meaningful for them.

Our beliefs bias the way we understand the world, and by recognizing these biases, we can become better thinkers, better researchers, and ultimately better believers. But the moment your brain embraces one belief, it becomes biased against any belief that contradicts the ones you hold. Thus, for every positive belief we have, there's a negative belief that can be either conscious or unconscious. For example, if you choose to believe in evolution and the Big Bang theory, your brain will automatically disbelieve biblical creationists who claim the earth was created in six days. Obviously, neither belief can be confirmed by our senses, but when we accept one statement as true, the brain automatically assumes that contradictory statements are false.

Our brains do not like ambiguity—a cognitive function called "uncertainty bias"—and as researchers at the University of California Los Angeles Brain Mapping Center found, it's regulated by the same frontal and parietal regions that are involved in Enlightenment experiences.[15] In other words, when you decrease your frontal lobe activity, as we've seen in the brain-scan studies reported in this book, your sense of certainty decreases. This makes it easier for the brain to engage in belief-changing activities.[16] But when the brain activity returns to normal after the experience, it reestablishes the sense of certainty of your new, enlightened beliefs in a powerful way.

Focusing on negative beliefs also has a deleterious effect on our emotions and our brain. In a study that included twelve thousand women, those who held negative beliefs about themselves were more at risk for developing depression, and those who had the least number of negative self-beliefs were the least likely to get depressed. Obviously, optimistic beliefs are healthier, but this requires us to ignore or interrupt the brain's natural propensity to ruminate on negativity.[17]

The next step in becoming a better believer is to recognize that the maps we build about the world can only approximate the truth. This is how I got so involved in my path toward Infinite Doubt, the experience I described in the first chapter. I realized there will always be a fundamental gap between our knowledge, our beliefs, and reality. In fact, our brain does not need absolute proof about anything in order to function and help us survive on a day-to-day basis. This is an important point to keep in mind when we examine our deepest beliefs in the quest for

Enlightenment: we don't necessarily have to be accurate about the world, only optimistic about the future. If you believe the world is a dangerous place, you'll blind yourself to the fact that over 90 percent of the world's population is kind, compassionate, and accepting of others. But the effect of that one "rotten apple" can spoil the entire barrel—a recurrent fact we see nearly every day in the news.

Becoming a better believer is a difficult task to undertake, for changing the brain requires patience and time. But if you succeed, even to some small degree, then you will be better able to embrace the beauty in this world. For this reason, I hold the deepest respect for those people who have had the courage to question and challenge their beliefs, for these are the individuals who have enriched our lives through their creativity and willingness to grow.

By questioning our own beliefs, and recognizing their limits, we open ourselves to change and to dissolving the imaginary boundaries that separate us from others. This, I would argue, is a path toward unity with others, and the transformation of beliefs is always one of the hallmarks of Enlightenment. As one of our survey respondents reported after a particularly powerful experience:

> This was not an effect that I had been consciously looking for at the time. And it was definitely a "conversion experience," a moment when my beliefs about myself and the world were radically shifted. It's been many years, and I remain transformed.

When it comes to changing our beliefs, we would expect to see corresponding changes in our brain. The data, while not definitive, strongly point to the fact that different beliefs fundamentally alter neural activity. For example, when I compared long-term meditators to nonmeditators, I found that the areas associated with belief formation were different. In fact, the more positive your beliefs become, the more pleasure you'll feel in your brain and your life.[18]

When frontal lobe activity goes down, old beliefs are interrupted as we surrender our conscious control to more intuitive processes in the brain. New neural connections are formed, giving rise to new ideas and beliefs. Enlightenment weakens old beliefs, and new beliefs will change the brain in order to help us envision reality in radically different ways.

As you go through the exercises in the next few chapters, think about your current beliefs and challenge them as you explore new ways to see yourself in this vast mysterious world, a world that your brain barely can begin to grasp. Seek Enlightenment—big or small—because the new beliefs you form will create more inner peace and greater cooperation with everyone you meet. And maybe, just maybe, as you venture from Plato's cave of ignorance, others might follow your path.

PART 3

MOVING TOWARD ENLIGHTENMENT

●

For those who seek Unity

There are many ways to pray,

And when I pray there is neither belief nor unbelief.

This body keeps me from the Beloved

Trapping me in the colors and scents of the world.

Tonight I will drown myself in the veils of happiness.

Oh God: Illuminate me with your perfume

And free me from this human wall of grief.

We are born from love

But I cannot dance without your melody—

There is so much work to do![1]

—*Rumi, thirteenth-century Sufi poet*

TEN

Preparing for Enlightenment

In Chapter 1, I described my journey toward Enlightenment, which I shared with Mark when we first teamed up to do research together. He shared some of his own mystical experiences, but my adventure seemed to trigger a unique transformational moment for him. I'll let Mark describe it in his own words.

MARK'S STORY

I clearly remember the day Andy described his experience of entering a state of Infinite Doubt. To me, it sounded excruciating, but Andy disagreed, saying that it freed him from years of painful questioning.

Personally, I couldn't imagine enjoying such a state, particularly since I had spent most of my post-adolescent years immersed in chronic self-doubt, but Andy's story reminded me of my own journey of self-discovery. Twenty-five years ago, when I was thirty-eight, I had what I would call a spontaneous mystical experience, similar to what many others described in Andy's spirituality survey. I was sitting in my office chair, gazing mindlessly out of the window. Suddenly, and for no reason that I could identify, I was filled with inner peace. I looked at the tree and it seemed "perfect." I looked at the fence and it seemed perfect. Even the weeds seemed perfect, and everything felt connected to everything else. I too felt perfect and connected to the tree, the fence, and the weeds. It was pure bliss and I clearly remember what I first said: "Oh! This is what those Buddhists and Hindus were writing about when they described enlightenment."

At that moment, my beliefs suddenly changed. I "knew" that there was no heaven or hell or god, and that when I died, that was the end of everything. Rather than feeling anxious or sad, I was filled with an incredible sense of being alive and living in the present moment, something I had never experienced before.

The feeling lasted for several months, but I slowly returned to my usual state of self-doubt. It did, however, inspire me to experiment with different forms of meditation and mindfulness—a practice where all you do is neutrally watch your feelings and thoughts flow through your mind. For me, the practice felt similar to the psychoanalytic technique known as free association, in which you just allow yourself to think about whatever ideas or words pop into your head.

As my feelings of doubt returned, I had a second unique experience that was very different from my first mystical "event," but equally transformational. At the time, I was deeply involved in spiritual and psychological research. Then one day, without any warning or preparation, I heard a small "voice" whisper to me:

"Mark, you don't know a damned thing about religion or psychology!"

I immediately—and inexplicably—felt a sense of joy, similar to what a prospector must feel like when he comes across a giant gold nugget. Instead of thinking that this was my old inner critic, I again "knew" that I had stumbled on a fundamental truth. I *knew* that I didn't know anything! Now Socrates seemed to be content when he made a similar statement at the end of his life, but my "aha" experience—and the sudden joy of self-illumination— quickly disappeared. I literally went from "Oh wow!" to "Oh crap! What do I do now?"

Here's the thing about mini-enlightenment experiences: self-discovery helps you to see the bigger picture, a larger truth, but it doesn't necessarily tell you what you should do next. And yet I instantly knew what I needed to do: study! So I started reading everything I could get my hands on concerning psychology, religion, spirituality, consciousness, and human nature, and this preoccupation changed the entire course of my life.

My obsession turned into an occupation as I started to review nearly three hundred books a year for several academic journals. I was so excited by what I was learning that I wanted to share it with others, and this led me to becoming a published author. In 2002, I teamed up with Andy, but I had had no other profound experiences since my "Oh wow! Oh crap!" experience in 1990. Instead, I simply continued experimenting with different meditations, capturing small glimpses of the first mystical experience that allowed me to feel a sense of unity with the world.

Andy's encounter with the sea of Infinite Doubt changed all that, and several months after he told me his story, I had one of the most amazing experiences in my life. I was watching a 1999 movie called *The Third Miracle*, about a sacrilegious priest hired by the Vatican to debunk false claims of miracles, something I could relate to because I am quite a skeptic when it comes to religious ideology.

The movie begins with a scene that takes place in a small European village during World War II. You can hear the sound of falling bombs, and then a young girl looks up to the sky and begins praying. Mysteriously, the sound of the bombs disappears and the village is saved. The movie's plot includes the investigation of the event, which is finally deemed to be a validated miracle.

I had seen the movie once before, but this time, after it ended, I stepped outside into the sunlight and found myself utterly free of any feelings of doubt—almost the opposite experience that Andy had had. My first thought was this: "Ah, this must be what others call 'grace.'"

I was almost overwhelmed by this absence of doubt because I had never experienced anything like it in my life. And yet it was the primary reason why I felt driven to read so many books: I was searching for inner peace. My last "aha" experience—where I realized I didn't know much about anything—had lasted only a few seconds, but this "doubt-less" state continued for nearly two months. I seemed to radiate a sense of well-being, and I was suddenly invited to give dozens of talks. It also established a new goal in my life: to find the most effective ways to elicit personal transformation, in myself and in the lives of others.

It also changed my beliefs about Enlightenment: instead of only seeking the big "E" experience, I find it highly valuable to savor the small "e" experiences that often occur in our lives. For me, each small experience gives me a great sense of meaning and helps to illuminate my purpose. I am satisfied with simply being on the path and less concerned about discovering ultimate truths. I also believe that my doubt-free experience would not have occurred if Andy had not shared his story with me, and it is my hope that the stories we've shared in this book will inspire you to consciously seek transformation.

DESIRE, SURRENDER, AND NEURAL RESONANCE

Mark brings up an important point: sharing your big "E" or small "e" moments can trigger similar experiences in other people. Based on research I've done for the past two decades, I'm convinced that the search for Enlightenment is hardwired in our brain. Perhaps because we are born with so little understanding of anything, we are conditioned to learn and grow, and as we change, our notions of reality continue to evolve.

When we hear people's experiences, observing their facial expressions and tone of voice, our brain begins to mirror the activity in the other person's brain. It's a well-documented phenomenon (but not specifically related to mirror neuron theory),[1] which is why we've described so many personal experiences from others. Reading or hearing about specific ideas or experiences helps your own brain to recall similar events in your past, and when you reflect on earlier transformations that improved the quality of your life, you stimulate your brain to seek more of them. Now there's no guarantee that any one exercise or meditation will bring you enlightenment, but our research has uncovered a few basic steps that can speed up the process:

1. First, you must genuinely *desire* insight and change, knowing that it could shake up some of your most cherished beliefs. Beliefs are principles that you formed in the past, and enlightenment—going by the dictionary definition—means "to bring new light to ignorance."

2. Second, you need to *prepare* yourself by engaging in gentle relaxation and awareness exercises. This will help prevent you from being overwhelmed by the next step.

3. Third, you'll need to *engage* in an intense ritual that will interrupt your old habits and everyday consciousness.

4. Fourth, you must completely *surrender* and immerse yourself in the ritual experience.

5. Finally, after you've completed your ritual, you must set aside ten to twenty minutes to deeply *reflect* on all of the feelings, thoughts, and sensations that occurred while you were in an altered state.

Desire—Prepare—Engage—Surrender—Reflect: These are the five basic steps that will prime your brain for Enlightenment.

Step 1 is crucial because it relates to the desire to change and the willingness to acknowledge doubt in the very beliefs you have held for most of your life. Are you ready and willing to accept an entirely different way of thinking? Are you willing to challenge your own beliefs—moral, political, or religious—and to be open to the possibility that your existing beliefs might be wrong? This is the state of mind that appears to be most conducive for triggering insights and the wide range of experiences associated with enlightenment. Step 1—the desire to change—may potentially protect you from feeling overwhelmed if and when an Enlightenment experience shatters your old worldview.

Step 2—preparation—is also essential because any form of physical or mental stress will stop you from entering into the creative and self-reflective stages described in our Spectrum of Human Awareness. That's why it is important to spend a few minutes relaxing your mind and all the muscles in your body. Yawning, slow focused breathing, and very slow stretching are the fastest ways to prepare for the next step.

Step 3—engaging in a ritual practice—can be done in many ways. All you need to do is create a repetitive movement or sound, or assume a specific posture (our research shows that it doesn't matter what you choose as long as it feels pleasant). However, the more unusual it is, the more your brain will become absorbed in the activity. This interrupts habitual forms of thinking and behavior that can stop you from entering into relaxed states of creativity and imagination.

Step 4—surrender—can be compared to what the esteemed psychologist Mihaly Csikszentmihalyi calls flow, a state of intense awareness where you become so immersed in an activity that your sense of self begins to disappear.[2] In that state, you can easily lose track of time, and so being in flow—surrendering to the experience itself without judgment or expectation—appears to be a precondition of enlightenment. Brain-scan research shows that there are significant drops of neural activity in the frontal lobe when you are in this state of flow.[3]

Step 5—deep reflection—is needed to integrate your experience back into your daily life. All you need to do is mindfully observe, without judgment, all of the feelings, thoughts, and sensations that naturally flow into consciousness. After ten or twenty minutes of reflection, ask yourself this question: "What insight—large or small—can I glean from this experience?" This helps to integrate the experience in ways that will change future behavior and beliefs.

Your everyday consciousness will initially resist any attempt to alter your sense of reality because it requires a lot of metabolic energy for the brain to accept and integrate new experiences, especially those that might change your beliefs and your perception of reality. Also, the human brain doesn't like ambiguity or surprises. It's willing to go after pleasurable experiences, but anything new and intense stimulates the danger circuits in your brain that are designed to shut down the higher states of consciousness, creativity, and imagination. But once you understand the process of resistance, it's easier to overcome. Here's where our Spectrum of Human Awareness can help.

The instinctual fear reaction takes place on Level 1, but as long as the environment seems safe, we'll spend most of our time in the areas I've circled in the illustration. Intentional "everyday" consciousness is in the center (Level 3), and it's the place from which

we are currently writing and from which you are currently reading or listening to this book. This is the language-centered awareness from which we make most of our intentional decisions to achieve specific goals.

To do this, the brain makes use of our memories and habituated behaviors that are part of Level 2. When we get exhausted, our mind begins to wander and daydream (Level 4). Although we're usually not aware of this creative process, the brain is actually solving problems as we relax. So, throughout most of the day, we spend our waking hours constantly shifting between Levels 2, 3, and 4.

But if we consciously decide to *observe* our feelings, thoughts, and fantasies that subliminally float through our awareness, we begin to shift into self-reflective states (Level 5) that are often described in spiritual practices. If we are fortunate enough to have an "aha" experience or major insight, our beliefs and behaviors can change (Level 6), and if the experience is strong enough, the qualities will be similar to the Enlightenment experiences we described in the earlier chapters of this book.

STEP 1: THE DESIRE TO CHANGE AND THE LIFE TRANSFORMATION INVENTORY

So how does one begin to *consciously* seek Enlightenment? The starting point is simple: if you want to transform the way you perceive yourself and the world, you must *intentionally* and passionately desire it. Conscious intentionality begins in your frontal lobe, and the thoughts and feelings embedded in this neurological process can influence many regions throughout the entire brain.[4]

Let's begin with an exercise that will "prime the pump" of Enlightenment by recalling past experiences that have transformed your life in positive ways. When we recall memories from the past, they contain remnants of the original experience, and if the experience was pleasurable, the motivational circuits in the brain will drive you toward other activities that could replicate the experience. For example, if you close your eyes and think about one of your favorite comfort foods, you'll probably salivate, and if you do so long enough, you'll find yourself moving toward the refrigerator door.

Enlightenment may not be that different. Here's what I'd like you to do. Visualize all of the past events that in some small or large way made your life feel more meaningful and purposeful. Think about books you have read that changed your outlook on life, or a teacher that showed you a side of yourself you had not recognized. Think about the people who have inspired you, or opened your heart, or taught you how to feel more connected to yourself and others. Deeply recall previous spiritual insights along with those "aha" moments that occurred while studying something new.

Take out a sheet of paper and write down as many of these life-changing events as you can. The writing is very important because if you only "think" your way through this exercise (which most people will do when reading or listening to a book or exercise), your mind will remain in its normal state of consciousness (Level

3). The writing process slows your thinking down, and as your eyes follow what you write, it becomes easier to shift into the creative and self-reflective states of consciousness (Levels 4 and 5) where Enlightenment begins.

First, take sixty seconds to yawn a few times and slowly stretch your arms, neck, and torso, becoming as relaxed as you possibly can. With pen in hand, respond to each of the topics below, and as you write, allow yourself to fully feel the memory as if you were living it for the very first time. If nothing comes to mind, move on to the next category:

* Books or movies that transformed or deeply affected your life.
* Teachers that deeply influenced your life.
* Childhood experiences that changed your view of yourself.
* Friendships that transformed some aspect of your life.
* Spiritual/religious experiences that "enlightened" you.
* Events/activities that changed your view of the world.
* Exceptional experiences that changed your relationship to money or work.
* Health-related experiences that dramatically improved your lifestyle.
* Realizations that profoundly changed your beliefs.
* Anything else that you would consider enlightening or life-transforming.

After you've listed your items, gaze at the page and allow yourself to daydream, immersing yourself in all of the feelings that flow through your semiconscious mind for five or ten minutes. Then consciously search for insights. Is there anything that "illuminates" you about your past, and does your list provide you with any new directions to take? Here is an example of how one of Mark's MBA students responded to this exercise:

> I realized that my life has been filled with many amazing moments, and I realized that I usually just recall the difficult times I've had. I immediately felt better about my whole life, my past and my present, and I felt more optimistic about the future. These feelings stayed with me for weeks!

Similar reports were given when we posted this exercise on a variety of social media websites. We strongly recommend you do this experiment with a family member or friend because you'll be surprised how many other positive memories come to mind as you share your insights and experiences. Again: when describing these positive events from the past, don't just talk *about* them; talk *from* them in a way that deeply resonates to the original experience. As the research shows, the more deeply you immerse yourself in these memorable moments, the more you'll feel a sense of well-being.[5]

However, the more you ruminate on negative memories or

worries from the past or about the future, the worse you'll feel. Even journaling about negative experiences can make you feel more emotionally distraught,[6] and the longer you write about them, the more anxious and depressed you'll become.[7] The reason for this is simple: your brain is designed to encode negative experiences into memory as a survival defense mechanism. It also has to work much harder to encode positive experiences, which is why I asked you to *immerse* yourself when recalling positive memories from the past.

As Barbara Fredrickson and other research psychologists discovered, if you want to build optimism and self-confidence, you have to maintain a "positivity ratio," where every negative feeling or thought needs to be offset by a minimum of three positive ones. If your positivity ratio falls below three to one, you're likely to be diagnosed with depression.[8] Fredrickson, a distinguished professor at the University of North Carolina, says that positivity is our birthright, and it is something we feel promotes the Enlightenment experience. She writes:

> Think of the times you feel connected to others and loved; when you feel playful, creative, or silly; when you feel blessed and at one with your surrounding; when your soul is stirred by the sheer beauty of existence; or when you feel energized and excited by a new idea or hobby. Positivity reigns whenever positive emotions—like love, joy, gratitude, serenity, interest, and inspiration—touch and open your heart.[9]

Barbara Fredrickson's research demonstrates that the more you accentuate the positive moments in your life—which is what the Life Transformation Inventory helps you to do—the more likely you'll have transformational experiences in your life. And if you push your positivity ratio up to five to one, the research shows that your business life will be more successful[10] and your personal relationships will flourish.[11] To me, that is truly an enlightening strategy.

STEP 2: PREPARING YOUR BODY AND MIND

The first step to changing everyday consciousness is to relax both your body and your mind, and the fastest way to physically relax is to move and stretch any part of your body—your head, arms, torso, etc.—in super-slow motion. The slower you go, the more you'll feel the subtle tensions in each muscle group. Try this simple experiment: Slowly rotate your head the way you normally do when you want to relax the muscles in your neck. Most people spend about five to ten seconds doing a single slow rotation. Now I want you to take a full sixty seconds to do the same activity. Notice all the tiny aches and pains? This super-slow motion interrupts your habitual way of relaxing, and it's the only way your brain can become aware of the tension you are holding. It is the awareness itself—a Level 5 process—and not the movement, that releases body tension.[12] Roll your head again in about five minutes and you will probably notice far fewer aches and pains.

Super-slow movement will increase frontal lobe activity, but the following exercise, which involves relaxing your mind, will cause it to drop (remember, the greater the change, the more powerful the experience). But how do you actually reduce brain "tension" that causes excessive activity, especially in the frontal lobe? Yawning, as we documented in *How God Changes Your Brain*, appears to be one of the fastest ways to eliminate neurological stress, and it dramatically slows down activity in the frontal lobe,[13] one of the steps that prepares your mind for Enlightenment.

Try this yawning exercise right now and notice how it changes your conscious awareness of your body and the environment. Begin by slowly yawning ten times, even if you don't feel like it. Fake the first few ones, making a sighing sound as you exhale, and soon you'll naturally begin to yawn. The more you yawn, the more you'll notice that your worries fade away, and you may even feel the first signs of euphoria, a clue that you are leaving the everyday consciousness of Level 3 and entering Level 4, where your mind begins to wander in a very spontaneous and creative way. If you can't yawn, don't fret. Just gently breathe in and out through your nose, as this will have similar effects on the brain.[14]

When you are mentally and physically relaxed, turn your attention to all of the subtle thoughts and feelings that are constantly flowing through your subconscious mind. Observe those thoughts, feelings, and sensations neutrally, without judging them. This is an important part of many spiritual contemplative practices, and it's when insights begin to occur. When a distracting thought floats into consciousness, just notice it, say "oh well," and watch it float away, bringing your attention back to your breathing and the sensations you are feeling in the present moment.

Here's another relaxation exercise you can try. Sit in a comfortable chair in a room where you won't be disturbed by other people or by phones. Open your mouth, breathe in slowly and gently, and as you breathe out, say "ah." This triggers a yawning and relaxation response, and as researchers at the University of Texas MD Anderson Cancer Center found, repeating "ah" improves cognitive function, mental health, and spiritual well-being.[15]

You should now feel more relaxed *and* alert, but more important, you'll find yourself in the present moment where most worrisome thoughts are suspended. To deepen your relaxation experience, begin to twist your torso in super-slow motion as you sit in your chair. Notice how each movement and stretch feels. This is a very effective way to interrupt negative self-talk, so try it next time you're being too hard on yourself about something and see what happens. Very slow movements also rapidly increase frontal lobe activity,[16] and if you do this before and after the more intense exercises described in the next chapter (which decrease neural activity), you'll generate more powerful shifts in consciousness.

STEPS 3 AND 4: CREATING A RITUAL AND SURRENDERING TO THE EXPERIENCE

Yawning, super-slow stretching, and focusing your awareness on the sensations happening in the present moment deepen any form of meditation, mindfulness, or spiritual practice, making it easier to have a transformational experience. While there is no set order for doing these steps, try to develop a program—a ritual—that works best for you to enter a deeply calm and restful state. This is particularly important before you delve into the more intense practices described in the next chapter.

The more relaxed you become, the easier it is to be fully immersed in the experience of the ritual. This is what we mean by "surrender." You let yourself go. You consciously interrupt the mind's propensity to analyze and interpret, and just let the experience you are having unfold. The more you immerse yourself in the sensations and feelings of relaxation, the more you will detach yourself from the everyday thoughts generated by everyday consciousness (Level 3).

You can spend as long as you'd like doing the exercises above, but the key is to do them only for as long as it feels comfortable and enjoyable. You don't have to push yourself to do an extended meditation or relaxation technique. In fact, we believe that one minute each hour, done throughout the day, may be just as effective as a single twenty-minute session, because the former integrates your spiritual practice into your workday, where it is most needed to reduce stress.

However, when you engage in the practices in the next chapter, we recommend that you spend ten to twenty minutes doing gentle relaxation before and after each intense meditation. If a particular technique or ritual becomes boring, change it. Boredom is a sign that an in-the-moment activity is becoming routine, reduced to a Level 2 habit that has little or no effect on your consciousness. This is what happened to Ibrahim, the Muslim we talked about in the previous chapter, who did Salat repetitiously.

STEP 5: REFLECTING ON YOUR EXPERIENCE

When you feel fully relaxed, allow your mind to wander and consciously pay attention to the partial fleeting thoughts, feelings, and images. Usually we have no idea where our "mind" goes, but the simple technique of mindfulness will show you how to become aware of a constant inner chatter operating in the back of your mind. We don't notice these "voices" because they're always there, shifting and changing so fast that it's almost impossible to track them.

Before I show you how to practice mindfulness, let me take a few moments to explain how different levels of consciousness generate different "voices" in the mind. Language is one of the most important tools for developing the human brain because it is the primary means we use to learn new information and to work with others to help us achieve our goals. By the time we are two years old, there's a constant dialogue occurring in our frontal lobe.[17] Although we are mostly unconscious of these thoughts, they help us make conscious decisions in the world by interrupting impulsive behavior.[18]

Here's the easiest way to become aware of this inner speech: close your eyes right now and try to remain free of any thought for the next sixty seconds. Most people can maintain inner silence for about ten seconds, and even advanced meditators rarely keep their minds quiet for longer than a minute. In fact, the harder you try to quiet your mind, the more you'll become aware of that cacophony of inner dialogues, opinions, and judgments that are generated by your frontal lobe. In essence, inner speech is the language of your creative mind—Level 4 in our Spectrum of Human Consciousness.[19] It's an ideal state to see the world in a different way.

But something else can happen as you observe the inner dialogues of the creative mind: you can begin to hear "other" voices, similar to what the Pentecostals and mediums experience when they engage in their unique communication strategies. These experiences—the "whispers" that you can notice when you enter Level 5 of mindful reflection—can usher in a new understanding about the world around you and dramatically alter your beliefs.

PRACTICING MINDFULNESS

Listening to your inner voices increases frontal lobe activity, but the exercises in the next chapter turn them off. In order to create the greatest neurological shift, we recommend doing mindfulness immediately before and after the intense practices we are about to describe. Mindfulness is conceptually easy; just sit back and *watch* all of the inner conversations that go on. Don't judge what you hear, just observe. By watching the productions of your mind, you develop what is often called an observing self. When you become aware of yourself watching yourself, you'll suddenly realize that you are not attached to any of the daily worries and frustrations that fill the day. Remember, though, that practicing mindfulness is sometimes hard to do. So if you feel you are struggling with it, give it time, don't judge yourself, and keep working on it.

In fact, the more you identify with your observing self, the more your confidence and self-esteem will increase. Why? Because the "observer" is a unique form of neurological awareness that occurs in a different part of your brain. It's not connected to the worrisome thoughts generated in the right frontal lobe, nor is it part of the optimistic thoughts generated by the left frontal lobe.[20] In other words, the more you add mindfulness to your daily routines, the easier it becomes to deal with many emotional problems. Self-reflective awareness reduces activity in the fear-and-worry centers of your brain,[21] and as you watch your anxiety, you become less anxious.[22] If you observe, without judgment, your depression, your blues begin to fade away.[23] Research also shows that for many people, mindfulness has transformed their lives by alleviating suffering from life-threatening diseases.[24]

Try this mindfulness exercise now. Close your eyes and go through the relaxation exercises described above. Now begin to watch your thoughts and feelings. You don't have to make them happen because they happen all the time. Notice how they continually change, and as you do so, yawn and stretch several times a minute to bring your mind back into a momentary state of silence. Here's a typical example of what goes on in your mind as you observe your thoughts and feelings without judgment:

1. Noticing my breath. Feels good. Wondering what will come next.
2. Heard a noise. Found myself distracted. Yawning feels good.
3. My back hurts. Nothing's happening. Am I doing this right?
4. I've got other things to do that are more important.
5. This really feels good. I'm bored. Ah, I'm beginning to get it!
6. Just aware. Calm. I'm not my thoughts.
7. Wow!

In this free-associative process, your frontal and parietal lobe activity will first increase. If you continue long enough, frontal and parietal activity may suddenly drop (in our studies it often took forty to sixty minutes before our subjects had this occur), but when that happens, most people feel a remarkable sense of peacefulness, clear-mindedness, flow, and unity—the hallmarks of Enlightenment.

Mindfulness is a slow path of transformation, but it trains your brain to remain calm when problems occur. In the techniques we'll introduce you to in the next chapter, neurological changes will take place rapidly, and if you do not have the foundations of mindfulness practice and the ability to remain deeply relaxed when you enter and exit altered states of consciousness, you can find yourself having experiences resembling a bad psychedelic trip. So start by adding relaxation and mindfulness to your repertoire of brain-training activities. It's your best insurance for getting the most out of the techniques described in the next chapter without stirring up a hornet's nest of negative feelings or memories, a common occurrence when people first engage in intense forms of chanting, meditation, and prayer.

Mindfulness keeps you and your brain grounded, bringing you back into the present moment where you can integrate your experience into the everyday consciousness that guides you through your daily tasks and interactions with others. To take you more deeply into these gentle mind-expanding techniques, we've created an eight-week audio training program called NeuroWisdom 101. It's described in detail in the appendix of this book and at www.NeuroWisdom.com. It contains many of the relaxation and positivity exercises we've documented in our previous books, along with fifty other experiential exercises that you can use to create your own ritual and personalized meditation practice.

ELEVEN

Intensifying the Experience

In our prior books we investigated specific brain changes related to gentle contemplative practices such as Buddhist meditation and Christian Centering Prayer. But Enlightenment isn't a *practice*, it's an emergent experience that can be triggered when the brain transitions from one stage of consciousness to the next. In both Eastern and Western philosophies, the big "E" experience is seen either as a gradual progression or ascension toward realization of the essential values identified by the various sacred texts or as a sudden burst of transformative insight often related to an intense spiritual practice.

Our research suggests that the more quickly you alter brain functioning by rapidly increasing and then decreasing neural activity, particularly in the frontal and parietal lobes, the more likely you are to interrupt old beliefs that could hold you back from experiencing life in a totally different way. When you take a plane trip, when do you feel the greatest change in the movement? It's not when you're sitting there on the tarmac, and it's not when you're actually cruising at five hundred miles an hour, thirty thousand feet above the ground. It's the moment of takeoff or landing—the incredible acceleration or deceleration—that dramatically wakes you up.

Now it doesn't seem to matter whether the process of Enlightenment is slow or fast, but when it does occur it can be spectacularly surprising, like when a supersonic jet suddenly breaks through the sound barrier. Boom! A sudden shift of consciousness takes place and everything changes in a way that brings insight, clarity, unity, peacefulness, or any of a dozen different states of illumination.

What we're beginning to see in these more unusual spiritual practices (Sufi chanting, speaking in tongues, psychography, etc.) is initially heightened activity throughout the brain followed by a sudden drop in which you often feel an incredible rush of energy and intensity. You've broken the "sound barrier" of everyday consciousness by moving into a new state of awareness.

Most researchers have focused on what happens in the brains of meditators while they are "cruising" along in the calmness of the experience and the changes that occur *after* they return to everyday reality. Even in our own surveys, we are recording people's experiences after the fact. But Enlightenment is the moment of transition, when our beliefs and worldview suddenly change in ways that provide us with new meaning, value, and purpose.

The brain-scan studies we've described in this book reveal some of the structures that underlie this moment of transformation. In the previous chapter, we gave you warm-up exercises to help provide the initial increase of activity in your brain—especially in the frontal and parietal regions. This helps prime the brain to experience different states of consciousness by eventually enabling a greater drop in activity in these areas that we believe are associated with Enlightenment experiences.

In this chapter, we'll guide you through practices that have been documented to dramatically turn off activity in those very same areas. By shutting down frontal lobe activity, your habits and belief system can be temporarily suspended, and by lowering activity in your parietal lobe, your ordinary sense of self can disappear from consciousness. In that moment, many people experience extraordinary insights that bring feelings of clarity, euphoria, and insights into the nature of themselves, the world, and the universe.

The power of Enlightenment isn't related just to any decrease in neural activity. Rather, it is the magnitude of the change that propels a person into these transformational states. By combining exercises in the previous chapter, which increase neural activity, and the exercises in this chapter, which lower it, you can metaphorically pull the rug out from under your normal way of processing the world.

Just the anticipation of having a euphoric or ecstatic state will release dopamine into the areas of your brain that regulate conscious experience, and the pleasure created will stimulate your brain to go deeper into exploring any activity associated with Enlightenment.[1] Our research, along with other studies, shows that having the conscious desire, intention, and commitment to seek Enlightenment helps trigger transformation. The same holds true in medicine: your intention and motivation to get better will actually trigger the healing processes in the body.[2] Of course, we can't guarantee that you will achieve Enlightenment since no one ever knows when or if that will happen. But our data support the approaches we describe in this book as ways of fostering an enlightenment experience.

A MEDICAL WORD OF ADVICE

Several years ago, when a television station interviewed me concerning meditation practices—which included a gentle breathing exercise—the producers wanted a medical disclaimer. Mark humorously suggested this: "Before breathing in or out, consult your doctor!" The literature does not show any risks for gentle breathing, but some meditation practices use forms of very deep and rapid breathing to trigger powerful shifts of consciousness. Even the exercises described in this chapter can quickly change the blood flow in your brain, so if you experience any dizziness or disorientation, please stop and check with your doctor. Although the risks are very low, for people who have severe emotional problems, psychoses, personality disorders, cognitive impairment, or cardiovascular problems, or are taking any medications relating to these symptoms, the following exercises should not be practiced without the guidance from a licensed health professional.[3]

TURNING SOMETHING INTO NOTHING

Our first exercise is one of the gentlest ways to lower activity in your frontal lobe. It's adapted from Zen meditation practice involving intense concentration. It's very effective for interrupting inner speech and making the sensations of the world around you fade away.[4] In this exercise you're going to make a sheet of paper disappear from the visual centers in your brain. You're going to turn "something" into "nothing."

In Zen practice, you are supposed to clear your mind of everything, which can be agonizingly difficult to do. The specific goal of Zen Buddhism is enlightenment, but most Zen teachers will tell you that any thoughts you have *about* enlightenment can potentially blind you to what it actually is. They don't want you to *think* about it, they want you to experience it with your senses:

A university professor went to visit a famous Zen monk to ask about enlightenment. The monk remained silent, but handed the teacher a teacup. The monk filled the cup to the brim but kept pouring. The professor cried out, "Stop! The cup is too full and I cannot sip the tea!" The monk then spoke: "Like this cup, your mind is filled with too many opinions and ideas. How can you taste enlightenment if you do not empty your cup?"

Do these anti-intellectual or "not thinking" approaches work? Apparently so. Brain-scan studies found that advanced Zen meditators can consistently lower frontal lobe activity in ways that lower their reactions to worries, fears, and doubts. They become more calm and serene—hallmarks of Enlightenment in the Zen tradition—and the studies even show that the size of their amygdala—a structure that helps regulate fear responses—had shrunk.[5] This change correlates to reductions in anxiety and stress and improved emotional regulation.

I'd like you to have the experience of emptying your mind—the teacup in the Zen tale—and turning off your consciousness for a few moments. If you can be aware of that state without thinking *about* it, you'll experience a very different emotional reality that feels exquisitely peaceful. All you need is a white sheet of paper. You can place it on your desk, or tape it to the wall, placing your chair in a way that you can comfortably look at it.

First, close your eyes, relax, and bring yourself into the present moment by paying attention to the natural rhythm of your breathing. Do this for several minutes, and then open your eyes and gaze at the paper using your fullest concentration. Explore it and study as many details as you can: the edge, the size, the whiteness of the paper, etc. Do this for two to three minutes.

Now begin to intently stare at the center of the white page. When your eyes wander to the edge, bring your attention back to the center. Keep gazing at the center until the edge of the paper disappears from consciousness. You won't see it anymore (a neural process known as selective inattention) and you'll cease to associate with the preconceived idea of paper. Consciousness changes into something undefinable as you literally decrease the brain's ability to think. Some people can do this within a few minutes, and for others, it can take much longer, but when it does occur, you might not actually be aware of anything, as if you were in a trance or a coma. The paper will have disappeared, along with your entire awareness of yourself.

For most people, the experience won't last long, but for a highly adept monk, another person might be needed to bring the practitioner back into everyday reality to interact with the world! Most people, however, will go back and forth between losing themselves in the whiteness and being jerked back into normal awareness. Neurologically, you are causing rapid increases and decreases in many areas of your brain, and this may trigger a sudden insight. Since it is easier to have an audio program guide you into this intense form of concentration, we've included it in the NeuroWisdom training described in the appendix. By consciously developing the ability to selectively eliminate other visual stimuli through this type of meditation practice, you can actually train yourself to become more of an optimist by ignoring the worrisome thoughts caused by increased activity in your frontal lobe.[6]

In the Sufi practice of Dhikr, some students will first gaze at a card with beautiful Arabic calligraphy depicting the word "Allah" before they enter into the more intense forms of chanting described below. Some Tibetans do a similar practice gazing at mandalas composed of geometric shapes and deities, and Navajo Indians use sand paintings to lose themselves in the spiritual symbols of their tradition. But gazing at the blank page is different. You're eliminating *all* of your ideas about everything—God, self, and especially Enlightenment—as you deliberately try to empty your mind of thoughts. When successful, the paper loses its meaning, and from that nothingness, something happens that is beyond words, and that, for the student of Zen, is Enlightenment. It also reflects a neuroscientific truth: your sense of the paper or yourself or anything else in the world is nothing more than a perceptual illusion, a construction that is created in your parietal lobe and then labeled in the language centers of your frontal lobe. In other words, our thoughts and feelings are just constructions in the mind—memories pulled from the past and projected onto the present moment. For many people, coming to this realization is a transformational insight.

AUTOMATIC WRITING

The Brazilian mediums in our study did a similar exercise to clear their minds before they channeled spirits of the dead. They closed their eyes and turned off the normal thinking processes associated with normal frontal lobe activity. As activity decreased, they entered trances, and as they wrote, they remembered nothing of what the "entity" was saying at the time. They just allowed their hands to move across the paper as words were written down.

So the question arises: Can anyone enter trance-like states to gain knowledge that is normally not accessible when we remain in a normal state of awareness? History certainly says yes, and doing automatic writing—where you allow a different "voice" to communicate to you or through you—turns out to be simple, fun, and surprisingly illuminating. It's as easy as asking an imaginary mentor to answer a question you are struggling with. For every person we can envision—a saint, an artist, a philosopher, an ancestor, etc.—our creative mind can speak to us through that imaginary muse. Such inner dialogues can have profoundly healing effects, helping us to solve problems that everyday consciousness cannot.[7] It's a form of channeling, and it really doesn't matter if you believe the voices are real or imaginary.

Channeling spiritual entities continues to be a popular activity in contemporary American culture. For example, Esther Hicks, a well-known author and public speaker, channels an enlightened entity named Abraham, seeing him as a source of inspiration and infinite intelligence. Hicks, like many other channelers, is comfortable with the notion that Abraham could simply be the voice of her own inner wisdom. JZ Knight, who has appeared often on television, communes with a thirty-thousand-year-old spirit named Ramtha, featured in the independent film *What the Bleep Do We Know?* Jane Roberts gained international fame in the 1970s by channeling an entity named Seth, and her books remain popular today.

Then there was Helen Schucman who, in 1965, began to receive dictation from an inner voice that identified itself as Jesus. The result was a best-selling book called *A Course in Miracles.* Dr. Schucman was a professor of medical psychology at Columbia University in New York, and for these reasons she chose to publish the material under a pseudonym. Carl Jung, the famous psychoanalyst (and Freud's closest associate), recorded many perceived encounters with two biblical entities, Elijah and Salome. He also had his own "inner guru" named Philemon and he used these spontaneous experiences to develop a technique called "active imagination," a form of conscious dreaming that Jung believed could help people uncover unconscious blocks and discover their true self.[8]

To practice channeling (which lowers activity in the frontal and parietal lobes), you'll need to have some paper and be in a quiet location where you won't be disturbed. First, decide what question you'd like to gain insight into, and write it down at the top of your page.

Now repeat the paper-gazing exercise I described above. Keeping the pen in your hand and your mind as "empty" as possible, see if you can allow another voice—a whisper, an intuition, a sensory impression—to come through to you. If nothing occurs, ask your question out loud. If you still don't hear anything, ask yourself, "What would _____ say?" and fill in the blank with someone—alive or dead—who you consider knowledgeable and wise. If your mind remains blank, you can rejoice in the fact that you just achieved the ability to have an empty mind! Or just use your imagination to make something up, because this will shift you out of the everyday consciousness of Level 3 and into the creative processes associated with Level 4.

Let's say you are struggling with a relationship. Write down as fast as you can, without deeply thinking about it, five different things you could do to make a tiny improvement. Use your imagination, be wild, and write whatever comes to mind. Ask yourself what the world's most famous psychologist might say to you. If you still have a problem, ask what your parents or friends would say—that too is a form of "channeling." If you really want to push the envelope, ask yourself what God or Jesus or Buddha would say, and write down everything that spontaneously comes to mind. Many people enjoy the exercise so much they can spend a half hour or longer channeling their intuition.

OM SWEET OM

Let's explore the power of a very simple sound meditation, one that will also lower frontal lobe activity and interrupt the inner speech that is a normal part of everyday consciousness. It involves the repetition of *OM*, often used in different Eastern meditation practices.

When researchers conducted an fMRI study on the OM meditation, they found that the longer people repeated the sound of *OM*, the less activity occurred in the frontal lobe, the anterior cingulate cortex, the thalamus, and the emotional centers in the limbic system.[9] The experience can be profound because everything you are consciously aware of begins to fade away: your thoughts, your feelings, even the sense of reality you have about the world surrounding you.

To test whether other sounds had the same effect, they had the same practitioners make an *ssss* sound, which eliminated the vibration sensations in the body and face. However, we really don't know why the *mmmm* part of the sound is so important for eliciting this decrease in neural activity, only that the *ssss* sound didn't affect the brain. The researchers went on to suggest that OM meditation could be beneficial in the treatment of depression and epilepsy, and we believe that it is also one of the easiest ways to prime the brain for Enlightenment.

For this exercise, all I want you to do is to sit upright in a chair and close your eyes. First, take sixty seconds to yawn, slowly stretch, and become as relaxed as you possibly can. Now take a very slow deep breath in through your *mouth,* and as you slowly exhale, make the sound of *oooohhhhmmmm,* drawing it out as long as you comfortably can. The *oh* sound will come out first, and as you slowly close your mouth, the *mmm* sound will vibrate your lips.

When the sound fades away, take another slow deep breath through your mouth. As you repeat this OM meditation, pay close attention to every nuance of the experience: the resonant sound, the sensations in your chest, throat, and face, and the tonal qualities as you say the *OM* sound in different ways. See if you can actually lose yourself in the sounds and sensations of *OM.*

Go as long as you like, and when your intuition tells you to stop, sit quietly for a few more minutes observing your feelings and thoughts. The OM meditation is even more powerful when you do it with a group of people because you'll quickly experience a sense of unity when all of the chanters' voices blend together. This is an example of what we call neural resonance, where the neurons in everyone's brain begin to fire in a similar way.

THE POWER OF REPETITION AND MOVEMENT

Several years ago, I conducted a brain-scan study with a group of cognitively impaired individuals. They performed a Kirtan Kriya meditation that involved melodically repeating the sounds *SA-TA-NA-MA*. At the same time, they touched each of their fingers with their thumbs. An audio recording with a guitarist and singer guided the participants through the twelve-minute exercise. The chanting was gentle, and we saw increases in the frontal and parietal areas similar to other contemplative meditations.[10] This, we believe, was the effect of the combination of repeating the sounds in the meditation along with the music and singing. As other researchers have demonstrated, just listening to pleasant music can increase frontal lobe activity and create states of bliss, so adding music to meditation can be very beneficial.[11]

But our Sufi practitioners did a far more forceful chant, rapidly repeating *lā 'ilāha 'illā-llāh* as they rhythmically rocked their heads and bodies. The chanting was monotone rather than melodic, and the sounds were uttered as the person deeply breathed in and out. Together all of these elements—increased breathing, repetitive sounds, rhythmic movement—created a powerful change in the brain. The result: frontal and parietal activity dropped as the meditation became more intense.

We also believe that adding movement to any meditation intensifies the experience, but there is very little research documenting the effects of rhythmic rocking. However, we do know that it is a normal activity for young children, helping them to coordinate the body with mental activity.[12] We know that gentle movement meditations like Tai Chi bestow many physiological and neurological benefits with virtually no side effects.[13] Movements done slowly increase activity in the frontal and parietal lobes, but faster rituals appear to rapidly decrease activity in these same areas, and our research with the Pentecostals and Sufis strongly suggests that fast rhythmic rocking and dancing can take you into ecstatic trance-like states in a matter of minutes.

Other spiritual traditions also believe that rhythmic activity takes you deeper into prayer and meditation. In the Jewish tradition, swaying the body (shuckling) during prayer can be traced back at least a thousand years, and many scholars argue that shuckling improves concentration and the emotional intensity of prayer.[14] In the Hasidic mystical tradition, some liken the practice of shuckling to the flame of a candle: it dances and sways to shake itself loose from the wick of your body so that your soul can reunite with the Source.

You now have a "formula" for consciously raising and lowering activity in different parts of your brain, and if you alternate between slow and fast movements, slow and rapid breathing, and reciting sounds in either a melodic or monotone way, you can create the greatest increases *and* decreases in neural activity. You can, at will, deepen your consciousness or obliterate it, stimulate your brain or calm it down, thus giving you more conscious control over your feelings, emotions, and thoughts.

THE POWER OF RITUAL

According to Harvey Whitehouse, the esteemed Oxford anthropologist, rituals are "the glue that holds social groups together."[15] Even microorganisms appear to exhibit ritual behaviors as a way to transmit information and knowledge.[16]

Almost all animals use complex movements, sounds, and other ritual behaviors to identify other members of their species and to mate. When these rhythmic activities kick in, their brain lights up and changes the way the animals look at the world. Rituals also create a primitive form of unity consciousness, where all of the brains in the group begin to fire in similar ways. They allow us to lose the distinction between self and other and feel an intense oneness or connectedness to everything, which would be Level 6 on our Spectrum of Human Awareness.[17]

In animal biology, many mating rituals contain *marked actions*: body gyrations and weird postures coupled with carefully stylized cries or songs. The behaviors serve a specific purpose by announcing the *intention* of the animal. The animal is saying, "I'm up to something special here, so don't misinterpret this as a threat."

We see the same elements in human ritual. Consider, for example, the marked actions used by priests, rabbis, imams, and shamans of all cultures—solemn bows, kneeling, prostration, ritualized walking with sacred texts, orchestrated movements of arms and hands, and the regular use of specific fragrances, food, and drink. Such actions bring our brain to attention and recognize that we are doing something important.[18]

In order for a ritual to help us reach a higher state of consciousness, it must contain rhythmic patterns that can be sustained over time, and virtually all religious traditions employ some kind of sustained rhythmic chant, prayer, hymn, or dance. These rhythms can be slow and stately, as in the praying of the rosary; lively and loud, as in the Pentecostal's speaking in tongues; or stirringly vigorous, as in the drum-driven Polynesian fertility rite.

It's no accident that human ceremonial rites—especially those relating to births, weddings, funerals, and adolescent rites of passage—tend to include elements of rhythm, repetition, and marked actions. In my view, this reflects a biological desire to form a united community where the you-versus-I mentality dissolves into oneness, acceptance, and cooperation. In essence, ritual helps us to shift between different levels of consciousness for the purpose of uniting with our highest spiritual and social principles.

THE MOST POWERFUL RITUAL IN THE WORLD?

If you want to quickly alter consciousness, intensify your ritual. The human brain naturally resists moving beyond the comfort zone of everyday consciousness, but many spiritual ceremonies overcome this resistance by including activities that will overload the senses: sitting in sweat lodges, engaging in drumming circles, taking mind-altering drugs, hyperventilating, fasting for several days, or performing painful rituals that will shock the body's nervous system. They can certainly elicit transformational experiences, but they can also traumatize you.

In order to avoid the emotional stress that can sometimes result from intense ritual practice, we suggest that you always begin with a series of relaxation exercises as we described in the previous chapter. Then spend a few minutes clarifying what your intention is. In other words, focus on the goal or objective you are aiming for.

You can start with any kind of ritual: a rhythmic movement of the body, the repetition of a meaningful word or phrase, reading scripture, counting your breath—anything that feels pleasant and meaningful for you. Intense concentration and immersion in the experience is essential for lowering brain activity, and the faster you go, the greater the neurological change. Also, the more rituals you can do together, the better. Try combining a sound or word with a body movement, and then try coordinating everything with your breathing. But don't go beyond your comfort zone.

It may take between ten to fifty minutes to reach a high level of intensity in which you might feel strong emotions or tingling sensations in different parts of your body. When you reach this point, begin to slow everything down: your breathing, your movements, your chanting, etc. You can even go super-slowly, noticing the tiny shifts in feelings and sensations. This will increase neural activity, and it is during this swing between high and low neural activity that Enlightenment experiences are most likely to occur. Don't spend more than an hour, and when you've finished your ritual, do another series of relaxation exercises.

Then allow yourself to observe all of the fragmentary thoughts and feelings that will begin to fill your consciousness. This is the most important step (Level 5, self-reflective awareness). This is the time when brain activity shifts quickly, when a sudden moment of insight might arise. It often helps to spontaneously write down any thoughts or insights you are having.

To summarize: first relax, then intentionally decide to seek Enlightenment. Find a meaningful word or phrase or sound and begin to repeat it. Add a simple body movement. Concentrate deeply on the sound and movement and slowly increase the speed. As you do so, your breathing will automatically deepen. Then surrender yourself to the experience. Continue until you feel a dramatic shift in awareness and then slow your movements and chanting down. Stop and slowly stretch and yawn, and observe the different thoughts and feelings that float into consciousness. Ask your intuition to provide you with an insight—small or large—about any aspect of your life and write down whatever thoughts come to mind.

CREATE YOUR PERSONAL DHIKR

Our research shows that the more you become familiar with foreign cultures and different spiritual traditions, the more tolerant and accepting you become of others *and* yourself. So I'd like to guide you through the elements of the Sufi ritual Dhikr, which literally means "remembrance" or "invocation," a process in which you immerse yourself in the hidden powers of the divine. We want to emphasize that Dhikr, like other rituals, can be personalized to help make it more meaningful. However, it should be realized that any time a ritual is personalized, you might not be adhering to the specific tradition anymore. Modifying the Dhikr ritual would not be fully consistent with Islam, but it might make it easier for you to utilize as part of your own path toward Enlightenment. Either way, this exercise will allow you to experience the beauty of this peaceful branch of mystical Islam.

Although you can do Dhikr by yourself, it is usually done with a group of people sitting or standing in a circle. Often there is a leader who stands in the center to set the pace for the ceremony, speeding up the chanting and slowing it down to achieve the desired state of spiritual awakening. Dhikrs can last less than an hour or continue for many hours, but for this experiment, I'll guide you through six rounds of chanting and rhythmic movement, each one lasting about five minutes. For advanced meditators, frontal and parietal activity can begin to drop within the first ten minutes, but for others it may take as long as an hour for this to happen.

Do these exercises sitting comfortably on a pillow on the floor, with enough space around you to sway your body to and fro. The faster you chant and move, the easier it becomes to slip into a trance-like state where you lose your sense of your body and your mind.

In Dhikr, the most common word that is repeated is "Allah." You pronounce it by drawing out the *ahh* and then forcefully saying *lahh* as you breathe out. You'll inhale quickly and then repeat the sequence. At first, it may feel a little awkward, but don't worry that you're doing it the wrong way because your body will automatically synchronize the sounds and breathing as you go faster.

If saying the word "Allah" bothers you, remember that this word is also used by Arab Christians to symbolize God and is derived from the ancient Hebrew terms of "il," "el," and "al."[19] In some mystical circles Allah means "breath" or "oneness." Even if it makes you uncomfortable, I recommend that you repeat this word for five minutes, focusing on the sounds of the syllables and the sensations it evokes in your body. If you still feel resistance, you could repeat the word "Isa" (pronounced "ee-sah," the Arabic name for Jesus) or "Musa" (pronounced "moo-sah," the name for Moses), or simply repeat the sound of *ahh*.

If you are willing, for just a few minutes, to surrender all of your religious beliefs and biases, we believe that this exercise will give you an extraordinary insight into yourself and the powerful practices of Sufism. Begin with Round 1, and then immerse yourself in each consecutive round.

ROUND 1: Sit in a chair or on the floor, in a comfortable position. Repeat *ahh-lahh* thirty to one hundred times. Use your intuition to slowly speed up and slow down the rate of your chanting. Spend at least five minutes doing this, and after you've finished, pause for sixty seconds and immerse yourself in the sensations of your body.

ROUND 2: This time, as you repeat *ahh-lahh*, rock your upper body forward with the *ahh* sound and backward as you say *lahh*. Use your intuition to slowly speed up and slow down your rocking and chanting. Do this for five minutes, and then remain silent and motionless for several minutes. You are training your brain to get used to consciously raising and lowering neural activity.

ROUND 3: This time I want you to repeat the two most sacred qualities of Allah: compassion and mercy. The Arabic words are "Rahman" (pronounced "rock-maan") and "Rahim" (pronounced "Raheem"). Here's what you repeat as you continue to rock back and forth: *Ir Rahman ir Rahim*, ("compassion and mercy") drawing out the vowel sounds of the second half of each word. Go through another thirty to one hundred repetitions, and after five to ten minutes, pause and savor the silence and the sensations in your body.

ROUND 4: Begin to chant *allah-hu* as you gently sway your body from side to side. "Hu" is another name for God and is loosely translated "God is." As you repeat this sound, notice how the chant begins to take on its own natural melody. This, by the way, is a documented neurological phenomenon and it explains why many people hear "celestial" music during certain meditations.[20] As you chant *allah-hu*, experiment with different ways of drawing out each syllable, immersing yourself in the patterns that sound the most beautiful. Repeat one hundred times or more (five to ten minutes) and then rest for sixty seconds, enjoying the aftereffects of this amazing experience.

ROUND 5: This round involves one of the most powerful Sufi chants, and it will take some practice to say it smoothly. You'll repeat this four-part phrase—*la-ilaha-il-allah*—in any way that makes it sound like a poetic phrase. It's pronounced like this: "laa-eee-lahh-haa il-ahh-lahh." But don't worry so much about the pronunciation; it's the rhythm and cadence that will put you into an ecstatic trance-like state.

As you say each part of the phrase—*la-ilaha-il-allah*—you'll move your head into a different position. With *la* gently tilt your head to the right. With *ilaha* bring your head to center and gently nod forward. Then bring your head back. When you say the next syllable—*il*—tilt your head to the left, and when you say *allah* bring your head back to center. Make the movements very small at first until you get the rhythm down and repeat the entire sequence for thirty to one hundred rounds, slowly speeding up for as long as you feel comfortable. You can do this as briefly as ten minutes, but the effects are far more powerful if you do it for twenty minutes or longer. Then slow down. When you have finished, close your eyes and ask the deepest parts of your being (or God) for a small insight into yourself or the nature of life. Don't be surprised if you hear a small "whisper" of wisdom coming through, and write down whatever you felt about the experience.

When creating your own version of Dhikr, don't worry if you are doing it "right." There are hundreds of American and European Sufi groups, and each one has their own version of these prayers. If you were to travel through North Africa and the Middle East, you'd encounter unique forms of movement, chanting, drumming, and singing in each tribal community. You can also go to YouTube and listen to a dozen different ways these prayers are performed.

Of course, if you believe in the fundamental principles of Islam, then this practice could have an even greater effect since you will be repeating phrases that your brain is deeply connected with. But regardless of your beliefs, make the sounds that feel natural to you and feel free to substitute different words or sounds as you create your own personalized ritual.

DANCE YOUR WAY TO ECSTASY

Of all the spiritual practices Mark and I have studied, and the many personal stories we have gathered from spiritual practitioners, the Sufi practice of Dhikr appears to be one of the more powerful strategies we've encountered for rapidly changing states of consciousness. It is similar to the Pentecostal practice of speaking in tongues, which uses gospel music and dancing as an induction into this brain-changing ritual. This raises an intriguing question: Will any form of rapid movement increase the likelihood of having enlightenment experiences? We think so, but only if you are consciously seeking a transformational insight. That appears to be a very important ingredient for Enlightenment.

In the 1960s many young people in the countercultural movement experimented with powerful drugs for the explicit purpose of discovering the true nature of consciousness. Many succeeded in changing their inner realities, but when psychedelics became an entertainment drug in the 1970s, few people described having any form of enlightenment experience. The same phenomenon occurred when MDMA, commonly known as Ecstasy, gained the reputation of being a therapeutic drug that could deepen intimacy between couples. Most people had breakthrough experiences, but when it became a party drug in the 1980s, that changed.

Drugs will definitely alter consciousness in radical ways, but when the search for enlightenment was no longer part of the equation, people just got high. They had fun, but they didn't have transformational experiences because they weren't looking for them. As we've emphasized, the *intention* increases the likelihood of transformation.

This is true for other ritual activities, like dancing, that can cause dramatic neurological changes in the brain. You can choose to do it for fun, or you can create a *ritual* with the clear intention of seeking Enlightenment. Here's a simple formula for using music and dance to alter your consciousness in ways that allow you to tap into the intuitive wisdom of your brain:

1. Make two lists of your favorite songs that create a state of bliss or euphoria. On one list include three to ten slow songs, and on the other list put down three to ten fast ones. Record each song on the list onto a player. You should now have between ten to fifty minutes of music.

2. Before you begin dancing to either list, state or write down, as clearly as possible, your Enlightenment goal. What problem would you like to gain insight into? What state of mind would

you like to experience (bliss, ecstasy, peace, clarity, etc.)? Or you can make an open-ended request to interrupt your everyday consciousness to see something new about yourself or the world.

3. Begin playing your slow list of songs first (you'll have to decide if songs with vocals are more distracting than instrumentals and tweak your list). When you first begin dancing, ritualize your movements, like a workout regime or a yoga agenda. Make each movement intentional and repetitive. Then play your fast list, keeping in mind your intention to erase your everyday consciousness and enter an altered state where you become one with the music. When you reach the point where you "lose yourself," you've entered the flow experience we described earlier in this book.

4. When you feel ready, slow down your dancing, turn off the music, and sit in silence, allowing your mind to spontaneously wander wherever it wants to go for a few minutes, then bring your attention back to the explicit goal of having an insight. Trust your intuition, and if something interesting comes to mind, write it down on a sheet of paper.

You can even do this in a group, as our Sufi participants and Pentecostal subjects often did. When everyone in the same room is committed to creating intentional change, remarkable things happen: personal problems seem to dissolve as everyone's brain resonates to the blissful state of one another. As researchers at the University Medical Center Groningen in the Netherlands discovered, dancing, chanting, and drumming together makes the entire group feel united.[21] This is a form of social and community enlightenment.

The same thing can occur during an exercise workout. Known as "runner's high," the rapid movement combined with the automatic deepening of breathing affects the structures of the brain in ways very similar to enlightenment experiences.[22] In fact, researchers have noted that a distinct "dissociation" takes place in the consciousness centers of high-performance athletes.[23] They become so totally immersed in the running experience that they enter a state of flow where nothing else exists in the world other than the experience itself. It's very similar to a hypnotic trance. Now, most runners and dancers do not report Enlightenment experiences during their exercise, but perhaps they would if they consciously *chose* to make enlightenment their goal as they engaged in vigorous exercise.

Research also shows that drumming is one of the fastest and safest ways to alter consciousness. When you rhythmically drum with others in a group, it neurologically bonds you to them in socially positive ways.[24] It reduces burnout at work[25] and it boosts your immune system.[26] Drumming even has similar effects on animals. For example, monkeys use drumming to communicate and coordinate behavioral responses,[27] birds will synchronize their head movements when they hear drumming,[28] and certain species of spiders will drum to turn on their mates.[29]

Group drumming has also been shown to help children stay focused and to gain better control over their emotions.[30] My suggestion: when a family problem occurs, add drumming to your dialogue. When you do so, discussing serious issues becomes fun and the entire family can solve problems with greater ease. As reported in the *American Journal of Public Health*, "drumming alleviates self-centeredness, isolation, and alienation, creating a sense of connectedness with self and others."[31] Drumming is a form of music and music is a form of communication, so why not sing, dance, and drum your way into social enlightenment and joy? It's one of the easiest and most powerful ways to change your brain.

EXERCISE YOUR WAY TOWARD ENLIGHTENMENT

The easiest and fastest way to increase activity in your frontal lobe above baseline (normal conscious activity relating to Levels 1 to 3 on the Spectrum of Human Awareness) is to engage in aerobic exercise. If you lie down with the intention of seeking Enlightenment, and deeply relax as you become aware of all of the different body sensations you are having (a meditation practice known as "yoga nidra"), frontal lobe activity quickly drops below baseline, moving you into Levels 4 and 5. It takes only three to five minutes of aerobics (example:

running in place as fast as you can) followed by five to ten minutes of restful awareness to trigger the neurological mechanisms that evoke sudden insight. But remember: you must have the *intention* to discover something new. According to researchers at the University of Georgia, going from rigorous exercise to a deep state of relaxation enhances brain development. This may, in fact, be the fastest way to learn how to prime your brain for Enlightenment by consciously shifting between different levels of consciousness.

RETURNING TO THE PRESENT MOMENT

Intense chanting and swaying can take you into realms that resemble powerful drug experiences, and for this reason, many spiritual leaders warn against staying in these altered states for longer than an hour. Historically speaking, ancient Jewish, Christian, and Islamic leaders often forbade any practice that generated ecstatic feelings since they believed it distracted a person from serious devotional prayer. Perhaps that is why, in our surveys, we found that people less aligned with traditional religion had more mystical and spiritual experiences.

But remember: when you engage in intense ritual practice, you run the risk of feeling disoriented for several hours or days. To avoid this side effect, we strongly advise you to practice the reflective calming exercises in the previous chapter after doing any exercises similar to those described above. The "reentry" process is simple: take several gentle breaths, then slowly stretch and yawn for at least sixty seconds. Then write, in a journal, a couple of paragraphs about your experience. Writing brings you back into the language centers of your brain that are turned off during these intensive practices. Here's another powerful technique that you can use to feel more grounded and serene. Make a list of your deepest values, and focus on one of them in the morning for sixty seconds. You can repeat the value word to yourself whenever you feel tired or irritable, and it will help to reduce stress and increase work productivity during the day.[32] Research from Harvard University found that repeating a single value word for twenty minutes will affect the function of twelve hundred stress-reducing genes.[33] Then whenever you feel stressed out, repeat this phrase putting one of your value words into the blank space: "I breathe in _____, I breathe out stress." Try it right now: repeat this phrase with any word you choose (love, confidence, peace, God, trust, etc.) and notice how it makes you feel.

In conclusion, it is the belief, intention, and implementation of your deepest values and desires that will transform an ordinary ritual into an extraordinary event. In that moment of enhanced awareness, enlightenment is possible . . . *but only if you ask!*

TWELVE

Enlightenment for All

Enlightenment—whether you define it as a spiritual awakening, a mystical experience, an intuitive insight, or a rational discovery—has the unique quality of transforming a person's perspective of reality in a fundamental way. As each of our brain-scan studies have shown, enlightenment can make permanent changes in the brain, and when this occurs, the way we perceive the world—and ourselves within it—also changes in dramatic ways. Old beliefs fall away and new ones are formed. New values become integrated into our personal relationships and work. Consciousness literally changes, and when that happens, reality seems different. You feel more alive, less attached to your worries, and more optimistic about the future. Even your ability to solve difficult problems becomes easier.

But what happens *after* you've had a transformational experience? Do you stay Enlightened or do you return to everyday consciousness? The answer, I believe, is best stated in the famous Zen story of a meditation student who diligently practiced day after day, month after month. Ten years passed, but he never reached Enlightenment. One afternoon, the teacher "saw" that his student was ready, and so he told the student that he must leave the monastery and climb to the top of a sacred mountain. There an enlightened Master lived, but the teacher did not tell this to his student.

The student, though brokenhearted over leaving his old community, resolved to climb to the top of the mountain and meditate there until he either reached Enlightenment or died. He ascended the slope, and when he was halfway to his destination, an old man carrying a large sack of laundry on his back was walking down from the mountaintop. This was the Master, and when he saw the young student climbing with such determination, he knew that the student was ready.

When the student looked up and gazed into the old man's eyes, the Master let go of his bag of laundry. The moment it hit the ground, the student became Enlightened, realizing he was in the presence of a great soul. He asked, "What now, Master?" But the Master said nothing. Instead, he slowly bent over, picked up his heavy bag, and placed it on his ancient shoulders. He then continued down the mountain and into the village, never to return again. The student understood.

Here is how I view the story of the Master and the student. For me, the bag of laundry symbolizes all of the old memories, feelings, thoughts, and beliefs that hinder us from seeing what reality actually is. We can sit all day in the monastic sanctuaries of the mind, but if we truly want to break free, we have to leave the comforts of our home—our everyday habits—and arduously struggle up the slope of consciousness in our search for deeper insights.

Perhaps we'll be lucky and experience Enlightenment before we reach the top of the mountain, perhaps not. But when that brief moment of illumination is sparked, all of our worries, fears, and doubts—our dirty laundry—fall from our shoulders. We momentarily bask in the experience, but then we must pick up our bag once more. After all, there's always more laundry to do: more work, more responsibilities, more challenges and struggles and pain. The difference is that Enlightenment puts all of these aspects of life into a fundamentally new perspective. And we maintain this new perspective permanently. It teaches us how everything, even our problems, is part of the journey toward greater consciousness and awareness. In other words, Enlightenment is that momentary glimpse of the thread that connects us to everyone and everything, and that awareness permanently changes the structure and functioning of our brain.

Of course you might be wondering, "This all sounds great, but is Enlightenment in my future? Is Enlightenment something I should strive for?" These are provocative questions to ponder, but because the brain is wired for change, I believe that Enlightenment is absolutely attainable by anyone. In fact, if the human brain is built to explore and understand our world, it would seem that the movement toward Enlightenment is an essential drive within every brain.

Indeed, the brains of today's scientists and educators are functioning in the same way as the ancient Egyptians, Akkadians, and Mayans: building new rituals and new psychologies and new perspectives to change the reality in which we live. We have amassed thousands of years of knowledge, and still we are not satisfied. We seek more of everything in our pursuit of knowing the truth. We have no choice because we are equipped with a transcendent brain, and in every culture throughout history, we see the perennial quest for Enlightenment.

This is the ultimate message of this book—Enlightenment is for anyone. The question becomes a matter of how to get there. Should we allow it to unfold naturally and spontaneously, or nudge it along through discipline, spiritual practices, or drugs? That is your personal choice, but no matter what form of Enlightenment you may seek—religious or secular—there seems to be a basic neurological rule: you must suspend, at least for a moment, every habituated belief you have previously held about yourself and the world. You have to interrupt the memories that constantly flood your everyday consciousness controlled by your frontal lobe and allow your existing belief system to break down. You'll experience the present moment with greater intensity as powerful neurochemicals are released throughout your brain. Your thoughts will change, your feelings will change, and new memories will be formed. It's not always easy, but your life, as you once knew it, will never be the same.

If you seek Enlightenment, expect some resistance because your ever-changing brain doesn't always like to change! It requires a lot of metabolic energy to alter the neuronal connections supporting old habits and behaviors, and so we consciously or unconsciously cling to our existing beliefs that had worked for us in the past. The brain is always balancing permanence versus change. While changes in the brain are inevitable and the drive toward the radical transformation related to Enlightenment are part of the brain's processes, there is always a competing desire to resist change, especially if it contradicts an old cherished, and useful, belief.

Any change in our environment, or in our brain, also stimulates the amygdala, causing us to react with uncertainty or fear. Enlightenment may be attainable, but it is equally laced with trepidation. This explains why, in our Belief Acceptance Scale, we found that most people are only partially open to new or different ideas. In fact, the stronger your belief system, the less likely you are to be open to other people who espouse different beliefs. But sometimes life's stressors are so strong that your belief system fails to relieve your suffering. In that deep despair, you may realize that it is time to try something new. And when you do, you'll influence everyone else in your life, starting with yourself.

Still, there is one more question: If Enlightenment can happen to anyone, can it also happen to everyone? There is every reason to think, given the similarities in all of our brains, that everyone can attain Enlightenment. If this were to happen, we can only imagine what this world would be like. If everyone viewed the world from his or her highest level of consciousness, we might expect to see an increased sense of compassion and openness to other people and other beliefs. This feeling of oneness and inclusiveness could be so pervasive that it might lead to a massive reduction in hatred and violence among people. While this may be incredibly optimistic, it was precisely the feeling I had those many years ago when I found myself immersed in that sea of Infinite Doubt. That experience wiped away my own pessimistic beliefs about humanity, leaving me filled with hope and optimism about our future.

I am reminded of one of my students who was contemplating her search for Enlightenment. She conjured up an unusual story about what was going to happen. "We are all like un-popped popcorn sitting in the bottom of a pot on a stove. Then the heat of transformation is turned up. One person will pop first, and just like a bag of popcorn, everyone else will be encouraged to pop." My student wanted to be Enlightened so that she could change the world, but she saw herself as a kernel waiting for the right moment to pop, unfold, and expand.

I loved her visual analogy, and that is why I have described as many different Enlightenment experiences as possible in this book, so that those stories would inspire you to pop. When someone describes their inner experience of unity, clarity, and transformation, those parts of your brain that regulate similar experiences are stimulated. Thus, by listening deeply to other people's Enlightenment stories, you are more likely to experience Enlightenment yourself. This, in turn, will encourage other people to deepen their consciousness and awareness.

Our data—the online spirituality survey, the Belief Acceptance Scale, and the numerous brain-scan studies we've described in this book—show that the human brain is primed for Enlightenment. All we need to do to unlock this process is find the right combination of practices, life experiences, and beliefs that will illuminate our path toward Enlightenment.

Enlightenment is a gift for all humanity. It is in your body and your brain, and it is waiting to be released in everyone.

APPENDIX

Tools and Resources to Foster Enlightenment

There is no magic pill, no drug, no single meditation or spiritual path that will guarantee Enlightenment. However, there are many books and gurus who promise transformation if you follow a specific formula or religion. Usually these resources are biased toward the underlying belief system of the teacher, and so we recommend that you use caution, especially when it comes to groups that employ powerful mind-altering techniques and meditations.

When exploring tools that enhance awareness, the first step is to identify what form of enlightenment you are seeking. Are you seeking spiritual forms of illumination? If so, check out online sites or organizations that offer guidance that aligns with your personal values and beliefs (for example, Christian Centering Prayer, the Spiritual Exercises of St. Ignatius, Jewish Hasidic practice, or local Sufi, Hindu, or Zen communities, etc.). In most religious groups, spiritual exercises are usually offered for free, so avoid organizations that demand large sums of money for their programs.

If you are seeking nonreligious avenues for enhancing serenity (the most common form of Enlightenment described in Eastern philosophies) or deep personal insight into yourself (the Western versions of Enlightenment), we suggest exploring the different forms of relaxation, stress-reduction, and mindfulness training classes offered at many universities. As we described in this book, these gentle exercises prepare the brain for the more intense changes of consciousness that take place when Enlightenment suddenly strikes.

Creating a daily personal practice—a ritual to exercise your brain—can increase the likelihood of having more insights and "aha" experiences, and so we created a self-guided audio training program called NeuroWisdom 101 that will guide you through a series of brief meditations that deepen relaxation, concentration, mindfulness, positivity, and emotional regulation. There are fifty-eight exercises in this program, and in Chapters 10 and 11, we take you through several of them.

NeuroWisdom 101 is also part of the executive MBA program at Loyola Marymount University, where it is used to reduce stress and boost work productivity. The mindfulness exercises included in NeuroWisdom 101 have been shown to improve sleep quality,[1] enhance immune functioning,[2] improve mood,[3] prevent work burnout,[4] increase job satisfaction,[5] and enhance empathy with others.[6]

You can find out more about this eight-week brain-training program by going to www.NeuroWisdom.com. We recommend that you use this program if you plan to engage in the more intense practices described in Chapter 11. For other tools and resources, visit www.AndrewNewberg.com or www.MarkRobertWaldman.com.

NOTES

Part 1. The Roots of Enlightenment

1 Original version and interpretation of Basho by Mark Robert Waldman, copyright 2014.

Chapter 1. The Enlightenment of a Troubled Kid

1 R. Descartes, *Meditations on First Philosophy: With Selections from the Objections and Replies*, trans. Michael Moriarty (Oxford, England: Oxford University Press, 2008).

Chapter 2. What Is Enlightenment?

1 Interpretive version by Mark Waldman, copyright 2014.

2 "The Mindfulness Revolution," *Time*, February 2014; see also Beth Gardiner, "Business Skills and Buddhist Mindfulness," *The Wall Street Journal*, April 3, 2012.

3 S. Hoeller, *Gnosticism: New Light on the Ancient Tradition of Inner Knowing* (Wheaton, IL: Quest Books, 2002).

4 Online Encyclopedia Britannica, 2014. http://www.britannica.com /topic/Manichaeism.

5 R. Porter, *The Enlightenment* (2nd ed.) (New York: Palgrave Macmillan, 2001).

6 D. Outram, *The Enlightenment: New Approaches to European History*, Vol. 3 (Cambridge, England: Cambridge University Press, 2013).

7 W. James, *The Varieties of Religious Experience*, Lecture 7 (London: Longmans Green and Co., 1902).

8 Online Encyclopedia Britannica: http://www.britannica.com
/EBchecked/topic/598700/Leo-Tolstoy/13426
/Conversion-and-religious-beliefs.

9 L. Tolstoy, *My Confession, My Religion: The Gospel in Brief* (New York: Thomas Y. Crowell Co., 1899), 63.

10 R. Bucke, *Cosmic Consciousness* (Philadelphia: Innes & Sons, 1901). (Note: Bucke wrote about his experience in third person; I took the liberty to put it into first person to improve its readability.—Mark Waldman.)

Chapter 3. What Enlightenment Feels Like

1 W. James, *The Varieties of Religious Experience* (London: Longmans Green and Co., 1902), 307.

Chapter 4. Enlightenment Without God

1 "Barna Study of Religious Change Since 1991 Shows Significant Changes by Faith Group," August 4, 2011. https://www.barna.org /barna-update/faith-spirituality/514-barna-study-of-religious -change-since-1991-shows-significant-changes-by-faith-group# .U9U47fldV8E; and Y. Anwar, "Americans and Religion Increasingly Parting Ways," UC Berkeley News Center, March 12, 2013.

2 Pew Research Center calculations based on the U.S. Census Bureau's August 2012 Current Population Survey, which estimates there are 234,787,000 adults in the U.S. http://www.pewforum.org /Unaffiliated/nones-on-the-rise.aspx%23growth.

3 "Three Spiritual Journeys of the Milliennials," The Barna Group, May 9, 2013. http://www.barna.org/teens-next-gen-articles /621-three-spiritual-journeys-of-millennials.

4 "Spirituality in Higher Education: A National Study of College Students' Search for Meaning and Purpose," 2003–2010, University of California, Los Angeles. http://spirituality.ucla.edu.

5 Speech to the German League of Human Rights, Berlin (Autumn 1932), as published in *Einstein: A Life in Science,* by Michael White and John Gribbin (New York: Free Press, 2005).

6 Published in *Albert Einstein: Philosopher-Scientist* (1949), ed. Paul A. Schilpp. Reprinted in *A Stubbornly Persistent Illusion: The Essential*

Scientific Works of Albert Einstein, ed. Stephen Hawking (Philadelphia: Running Press, 2009).

7 Parrott AC. Human psychobiology of MDMA or "Ecstasy": an overview of 25 years of empirical research. Hum Psychopharmacol. 2013; 28(4):289–307.

8 Smith DE, Raswyck GE, Davidson LD. From Hofmann to the Haight Ashbury, and into the future: the past and potential of lysergic acid diethlyamide. J Psychoactive Drugs. 2014;46(1):3–10; Krebs TS, Johansen PØ. Lysergic acid diethylamide (LSD) for alcoholism: meta-analysis of randomized controlled trials. J Psychopharmacol. 2012;26(7):994–1002; and Barbosa PC, Mizumoto S, Bogenschutz MP, Strassman RJ. Health status of ayahuasca users. Drug Test Anal. 2012;4(7-8):601–9.

9 Griffiths RR, Richards WA, McCann U, Jesse R. Psilocybin can occasion mystical-type experiences having substantial and sustained personal meaning and spiritual significance. Psychopharmacology (Berl). 2006;187(3):268–83.

10 Griffiths R, Richards W, Johnson M, McCann U, Jesse R. Mystical-type experiences occasioned by psilocybin mediate the attribution of personal meaning and spiritual significance 14 months later. J Psychopharmacol. 2008;22(6):621–32.

11 Cummins C, Lyke J. Peak experiences of psilocybin users and non-users. J Psychoactive Drugs. 2013;45(2):189–94.

12 van Amsterdam J, Opperhuizen A, van den Brink W. Harm potential of magic mushroom use: a review. Regul Toxicol Pharmacol. 2011;59(3):423–9; and Hermle L, Kovar KA, Hewer W, Ruchsow M. Hallucinogen-induced psychological disorders. Fortschr Neurol Psychiatr. 2008;76(6):334–42.

Chapter 5. The Spectrum of Human Awareness

1 Zhao Q, Zhou Z, Xu H, Chen S, Xu F, Fan W, Han L. Dynamic neural network of insight: a functional magnetic resonance imaging study on solving Chinese "chengyu" riddles. PLoS One. 2013;8(3):e59351.

2 Pew Forum on Religion and Public Life, "Global Christianity: A Report on the Size and Distribution of the World's Christian Population," December 19, 2011.

3 Qiu J, Li H, Jou J, Liu J, Luo Y, Feng T, Wu Z, Zhang Q. Neural correlates of the "Aha" experiences: evidence from an fMRI study of insight problem solving. Cortex. 2010;46(3):397–403.

4 Newberg A, Alavi A, Baime M, Pourdehnad M, Santanna J, d'Aquili E. The measurement of regional cerebral blood flow during the complex cognitive task of meditation: a preliminary SPECT study. Psychiatry Res. 2001;106(2):113–22.

5 Antonio Damasio, in his book *Self Comes to Mind* (New York: Pantheon, 2010), refers to this quality of having a sense of "self." First we are awake, and this primal awareness generates our cognitive decision making, or what Damasio calls the "mind." Thus, awareness plus mind plus self equals consciousness, a model that relies heavily on the neurobiological research of Jaak Panksepp.

6 Wallis LJ, Range F, Müller CA, Serisier S, Huber L, Zsó V. Lifespan development of attentiveness in domestic dogs: drawing parallels with humans. Front Psychol. 2014;5:71.

7 Shettleworth SJ. Do animals have insight, and what is insight anyway? Can J Exp Psychol. 2012;66(4):217–26.

8 Drayton LA, Santos LR. A decade of theory of mind research on Cayo Santiago: Insights into rhesus macaque social cognition. Am J Primatol. 2014; doi:10.1002/ajp.22362.

9 Nomura T, Murakami Y, Gotoh H, Ono K. Reconstruction of ancestral brains: Exploring the evolutionary process of encephalization in amniotes. Neurosci Res. 2014. pii: S0168-0102(14)00041-8. doi:10.1016/j.neures.2014.03.004. Frasnelli E. Brain and behavioral lateralization in invertebrates. Front Psychol. 2013;4:939. Aru J, Bachmann T, Singer W, Melloni L. Distilling the neural correlates of consciousness. Neurosci Biobehav Rev. 2012;36(2):737–46.

10 Wei D, Yang J, Li W, Wang K, Zhang Q, Qiu J. Increased resting functional connectivity of the medial prefrontal cortex in creativity by means of cognitive stimulation. Cortex. 2014;51:92–102.

11 Silberstein RB, Nield GE. Measuring emotion in advertising research: prefrontal brain activity. IEEE Pulse. 2012 May–Jun;3(3):24–7; and Ding X, Tang YY, Tang R, Posner MI. Improving creativity performance by short-term meditation. Behav Brain Funct. 2014;10:9.

12 Ding X, Tang YY, Cao C, Deng Y, Wang Y, Xin X, Posner MI. Short-term meditation modulates brain activity of insight evoked with solution cue. Soc Cogn Affect Neurosci. 2014; 10(1):43–9.

13 Bhasin MK, Dusek JA, Chang BH, Joseph MG, Denninger JW, Fricchione GL, Benson H, Libermann TA. Relaxation response induces temporal transcriptome changes in energy metabolism, insulin secretion and inflammatory pathways. PLoS One. 2013 May 1;8(5):e62817.

14 Qiu J, Li H, Jou J, Liu J, Luo Y, Feng T, Wu Z, Zhang Q. Neural correlates of the "Aha" experiences: evidence from an fMRI study of insight problem solving. Cortex. 2010;46(3):397–403.

15 Zhao Q, Li Y, Shang X, Zhou Z, Han L. Uniformity and nonuniformity of neural activities correlated to different insight problem solving. Neuroscience. 2014;270:203–11. Aziz-Zadeh L, Kaplan JT, Iacoboni M. "Aha!": The neural correlates of verbal insight solutions. Hum Brain Mapp. 2009;30(3):908–16.

16 Innes KE, Selfe TK. Meditation as a therapeutic intervention for adults at risk for Alzheimer's disease—potential benefits and underlying mechanisms. Front Psychiatry. 2014;5:40; and Marciniak R, Sheardova K, Cermáková P, Hudeček D, Sumec R, Hort J. Effect of meditation on cognitive functions in context of aging and neurodegenerative diseases. Front Behav Neurosci. 2014;8:17.

17 Kounios J, Frymiare JL, Bowden EM, Fleck JI, Subramaniam K, Parrish TB, Jung-Beeman M. The prepared mind: neural activity prior to problem presentation predicts subsequent solution by sudden insight. Psychol Sci. 2006;17(10):882–90.

Part 2. The Paths Toward Enlightenment

1 William Blake (1757–1827), "The Auguries of Innocence," in *The Harvard Classics—English Poetry, Volume II: From Collins to Fitzgerald*, ed. C. Eliot (Collier, 1910), 356.

Chapter 6. Channeling Supernatural Entities

1 Avenary H. The Hasidic Nigun. Ethos and Melos of a folk liturgy. J International Folk Music Council. 1964(16):60–63.

2 Newberg AB, Wintering NA, Morgan D, Waldman MR. The measurement of regional cerebral blood flow during glossolalia: a preliminary SPECT study. Psychiatry Res: Neuroimaging. 2006;148(1):67–71.

3 R. Forbes, "Slavery and the Evangelical Enlightenment," in *Religion and the Antebellum Debate over Slavery*, eds. McKivigan and Snay (Athens: University of Georgia Press, 1998).

4 Qiu J, Li H, Jou J, Liu J, Luo Y, Feng T, Wu Z, Zhang Q. Neural correlates of the "Aha" experiences: evidence from an fMRI study of insight problem solving. Cortex. 2010;46(3):397–403.

5 Beischel J, Schwartz GE. Anomalous information reception by research mediums demonstrated using a novel triple-blind protocol. Explore (NY). 2007;3(1):23–7.

6 Moreira-Almeida A, Neto FL, Cardeña E. Comparison of Brazilian spiritist mediumship and dissociative identity disorder. J Nerv Ment Dis. 2008;196(5):420–4; and Moreira-Almeida A, Lotufo Neto F, Greyson B. Dissociative and psychotic experiences in Brazilian spiritist mediums. Psychother Psychosom. 2007;76(1):57–8.

7 Shenefelt PD. Ideomotor signaling: from divining spiritual messages to discerning subconscious answers during hypnosis and hypnoanalysis, a historical perspective. Am J Clin Hypn. 2011;53(3):157–67.

8 Faymonville ME, Boly M, Laureys S. Functional neuroanatomy of the hypnotic state. J Physiol Paris. 2006;99(4–6):463–9.

9 Peres JF, Moreira-Almeida A, Caixeta L, Leao F, Newberg A. Neuroimaging during trance state: a contribution to the study of dissociation. PLoS One. 2012;7(11):e49360.

10 Liu S, Chow HM, Xu Y, Erkkinen MG, Swett KE, Eagle MW, Rizik-Baer DA, Braun AR. Neural correlates of lyrical improvisation: an FMRI study of freestyle rap. Sci Rep. 2012;2:834.

11 Abuhamdeh S, Csikszentmihalyi M. The importance of challenge for the enjoyment of intrinsically motivated, goal-directed activities. Pers Soc Psychol Bull. 2012;38(3):317–30.

12 Kirchner JM. Incorporating flow into practice and performance. Work. 2011;40(3):289–96; and Seligman ME, Csikszentmihalyi M. Positive psychology. An introduction. Am Psychol. 2000;55(1):5–14.

13 Furnes B, Dysvik E. A systematic writing program as a tool in the grief process: Part 1. Patient Prefer Adherence. 2010;4:425–31.

14 Lepore SJ. Expressive writing moderates the relation between intrusive thoughts and depressive symptoms. J Pers Soc Psychol. 1997;73(5):1030–7.

15 Lotze M, Erhard K, Neumann N, Eickhoff SB, Langner R. Neural correlates of verbal creativity: differences in resting-state functional connectivity associated with expertise in creative writing. Front Hum Neurosci. 2014;8:516.

Chapter 7. Changing the Consciousness of Others

1 Wu Jiang, *Enlightenment in Dispute* (Oxford, England: Oxford University Press, 2008).

2 Harris WS, Gowda M, Kolb JW, Strychacz CP, Vacek JL, Jones PG, Forker A, O'Keefe JH, McCallister BD. A randomized, controlled trial of the effects of remote, intercessory prayer on outcomes in patients admitted to the coronary care unit. Arch Intern Med. 1999;159(19):2273–8; and Byrd RC. Positive therapeutic effects of intercessory prayer in a coronary care unit population. South Med J. 1988;81(7):826–9.

3 Matthews DA, Marlowe SM, MacNutt FS. Effects of intercessory prayer on patients with rheumatoid arthritis. South Med J. 2000;93(12): 1177–86.

4 Aviles JM, Whelan SE, Hernke DA, Williams BA, Kenny KE, O'Fallon WM, Kopecky SL. Intercessory prayer and cardiovascular disease progression in a coronary care unit population: a randomized controlled trial. Mayo Clin Proc. 2001;76(12):1192–8; and Matthews WJ, Conti JM, Sireci SG. The effects of intercessory prayer, positive visualization, and expectancy on the well-being of kidney dialysis patients. Altern Ther Health Med. 2001;7(5):42–52.

5 Walker SR, Tonigan JS, Miller WR, Corner S, Kahlich L. Intercessory prayer in the treatment of alcohol abuse and dependence: a pilot investigation. Altern Ther Health Med. 1997;3(6):79–86.

6 Mathai J, Bourne A. Pilot study investigating the effect of intercessory prayer in the treatment of child psychiatric disorders. Australas Psychiatry. 2004;12(4):386–9.

7 Astin JA, Stone J, Abrams DI, Moore DH, Couey P, Buscemi R, Targ E. The efficacy of distant healing for human immunodeficiency virus—results of a randomized trial. Altern Ther Health Med. 2006;12(6):36–41.

8 Schlitz M, Hopf HW, Eskenazi L, Vieten C, Radin D. Distant healing of surgical wounds: an exploratory study. Explore (NY). 2012;8(4):223–30.

9 da Rosa MI, Silva FR, Silva BR, Costa LC, Bergamo AM, Silva NC, Medeiros LR, Battisti ID, Azevedo R. A randomized clinical trial on the effects of remote intercessory prayer in the adverse outcomes of pregnancies. Cien Saude Colet. 2013;18(8):2379–84.

10 Nelson R, Bancel P. Effects of mass consciousness: changes in random data during global events. Explore (NY). 2011;7(6):373–83.

11 Benson H, Dusek JA, Sherwood JB, Lam P, Bethea CF, Carpenter W, Levitsky S, Hill PC, Clem DW Jr, Jain MK, Drumel D, Kopecky SL, Mueller PS, Marek D, Rollins S, Hibberd PL. Study of the Therapeutic Effects of Intercessory Prayer (STEP) in cardiac bypass patients: a multicenter randomized trial of uncertainty and certainty of receiving intercessory prayer. Am Heart J. 2006;151(4):934–42.

12 Palmer RF, Katerndahl D, Morgan-Kidd J. A randomized trial of the effects of remote intercessory prayer: interactions with personal beliefs on problem-specific outcomes and functional status. J Altern Complement Med. 2004;10(3):438–48.

13 Hefti R, Koenig HG. Prayers for patients with internal and cardiological diseases—an applicable therapeutic method? MMW Fortschr Med. 2007;149(51–52):31–4.

14 Moreira-Almeida A, Koss-Chioino JD. Recognition and treatment of psychotic symptoms: spiritists compared to mental health professionals in Puerto Rico and Brazil. Psychiatry. 2009;72(3):268–83.

15 Hoşrik EM, Cüceloğlu AE, Erpolat S. Therapeutic effects of Islamic intercessory prayer on warts. J Relig Health. 2014; 10.1007/s10943-014-9837.

16 Schjoedt U, Stødkilde-Jørgensen H, Geertz AW, Lund TE, Roepstorff A. The power of charisma—perceived charisma inhibits the frontal executive network of believers in intercessory prayer. Soc Cogn Affect Neurosci. 2011;6(1):119–27.

17 Rouder JN, Morey RD, Province JM. A Bayes factor meta-analysis of recent extrasensory perception experiments: comment on Storm, Tressoldi, and Di Risio (2010). Psychol Bull. 2013;139(1):241–7.

18 Mossbridge JA, Tressoldi P, Utts J, Ives JA, Radin D, Jonas WB. Predicting the unpredictable: critical analysis and practical implications of predictive anticipatory activity. Front Hum Neurosci. 2014;8:146.

19 Kuo WJ, Sjöström T, Chen YP, Wang YH, Huang CY. Intuition and deliberation: two systems for strategizing in the brain. Science. 2009 Apr 24;324(5926):519–22.

20 Allman JM, Watson KK, Tetreault NA, Hakeem AY. Intuition and autism: a possible role for Von Economo neurons. Trends Cogn Sci. 2005;9(8):367–73.

21 Hsu M, Anen C, Quartz SR. The right and the good: distributive justice and neural encoding of equity and efficiency. Science. 2008;320(5879):1092–5.

22 Ralph Oesper, *The Human Side of Scientists* (Cincinnati, OH: University Publications, 1975).

23 Błażek M, Kaźmierczak M, Besta T. Sense of purpose in life and escape from self as the predictors of quality of life in clinical samples. J Relig Health. 2015; 54(2):517–23.

24 Mariano JM, Going J. Youth purpose and positive youth development. Adv Child Dev Behav. 2011;41:39–68.

25 Keng SL, Smoski MJ, Robins CJ. Effects of mindfulness on psychological health: a review of empirical studies. Clin Psychol Rev. 2011;31(6):1041–56.

26 Potter PJ. Energy therapies in advanced practice oncology: an evidence-informed practice approach. J Adv Pract Oncol. 2013;4(3):139–51.

Chapter 8. Opening the Heart to Unity

1 Sauvage C, Jissendi P, Seignan S, Manto M, Habas C. Brain areas involved in the control of speed during a motor sequence of the foot: real movement versus mental imagery. J Neuroradiol. 2013;40(4): 267–80; Thaut MH, Demartin M, Sanes JN. Brain networks for integrative rhythm formation. PLoS One. 2008;3(5):e2312; and Esposti R, Cavallari P, Baldissera F. Feedback control of the limbs position during voluntary rhythmic oscillation. Biol Cybern. 2007;97(2):123–36.

2 Online Encyclopedia Britannica: http://www.britannica.com /EBchecked/topic/123937/Codex-Argenteus.

3 M. Smith, *The Origins of Biblical Monotheism: Israel's Polytheistic Background and the Ugaritic Texts* (Oxford, England: Oxford University Press, 2001).

4 W. A. Meyer, B. Hyde, F. Muqaddam, and S. Kahn, *Physicians of the Heart: A Sufi View of the 99 Names of Allah* (Sufi Ruhaniat International, 2011).

5 Gerard McCool, "The Christian Wisdom Tradition and Enlightenment Reason," in *Examining the Catholic Intellectual Tradition*, eds. Anthony Cernera and Oliver Morgan (Fairfield, CT: Sacred Heart University Press, 2000).

6 Muhammad Hozien and Valerie Turner, eds., *Al-Ghazali: The Marvels of the Heart*, Book 21 (Louisville, KY: Fons Vitae, 2010).

7 W. M. Watt, *The Faith and Practice of Al-Ghazali* (London: George Allen and Unwin Ltd., 1953).

8 Weng HY, Fox AS, Shackman AJ, Stodola DE, Caldwell JZ, Olson MC, Rogers GM, Davidson RJ. Compassion training alters altruism and neural responses to suffering. Psychol Sci. 2013;24(7):1171–80.

9 Light SN, Heller AS, Johnstone T, Kolden GG, Peterson MJ, Kalin NH, Davidson RJ. Reduced right ventrolateral prefrontal cortex activity while inhibiting positive affect is associated with improvement in hedonic capacity after 8 weeks of antidepressant treatment in major depressive disorder. Biol Psychiatry. 2011;70(10):962–8.

10 Doufesh H, Faisal T, Lim KS, Ibrahim F. EEG spectral analysis on Muslim prayers. Appl Psychophysiol Biofeedback. 2012;37(1):11–8.

11 Alabdulwahab SS, Kachanathu SJ, Oluseye K. Physical activity associated with prayer regimes improves standing dynamic balance of healthy people. J Phys Ther Sci. 2013;25(12):1565–8.

12 Hosseini M, Salehi A, Fallahi Khoshknab M, Rokofian A, Davidson PM. The effect of a preoperative spiritual/religious intervention on anxiety in Shia Muslim patients undergoing coronary artery bypass graft surgery: a randomized controlled trial. J Holist Nurs. 2013;31(3):164–72.

13 Moss AS, Wintering N, Roggenkamp H, Khalsa DS, Waldman MR, Monti D, Newberg AB. Effects of an 8-week meditation program on mood and anxiety in patients with memory loss. J Altern Complement Med. 2012;18(1):48–53.

14 Hansen G. Schizophrenia or spiritual crisis? On "raising the kundalini" and its diagnostic classification. Ugeskr Laeger. 1995;157(31):4360–2.

15 Wise RA. Dopamine and reward: the anhedonia hypothesis 30 years on. Neurotox Res. 2008;14(2–3):169–83.

16 von Kirchenheim C, Persinger MA. Time distortion—a comparison of hypnotic induction and progressive relaxation procedures: a brief communication. Int J Clin Exp Hypn. 1991;39(2):63–6.

17 Ulrich M, Keller J, Hoenig K, Waller C, Grön G. Neural correlates of experimentally induced flow experiences. Neuroimage. 2014;86:194–202.

18 Kounios J, Beeman M. The cognitive neuroscience of insight. Annu Rev Psychol. 2014;65:71–93.

19 Yoder KJ, Decety J. The Good, the bad, and the just: justice sensitivity predicts neural response during moral evaluation of actions performed by others. J Neurosci. 2014;34(12):4161–6.

20 Original version and interpretation of a Hafez poem by Mark Robert Waldman. Copyright 2010.

Chapter 9. Believing in Transformation

1 Heiphetz L, Spelke ES, Harris PL, Banaji MR. The development of reasoning about beliefs: Fact, preference, and ideology. J Exp Soc Psychol. 2013;49(3):559–65.

2 Epley N, Converse BA, Delbosc A, Monteleone GA, Cacioppo JT. Believers' estimates of God's beliefs are more egocentric than estimates of other people's beliefs. Proc Natl Acad Sci U S A. 2009;106(51):21533–8.

3 Ogawa A, Yamazaki Y, Ueno K, Cheng K, Iriki A. Neural correlates of species-typical illogical cognitive bias in human inference. J Cogn Neurosci. 2010;22(9):2120–30.

4 Galante J, Galante I, Bekkers MJ, Gallacher J. Effect of kindness-based meditation on health and well-being: A systematic review and meta-analysis. J Consult Clin Psychol. 2014; 82(6):1101–14.

5 Garrison KA, Scheinost D, Constable RT, Brewer JA. BOLD signal and functional connectivity associated with loving kindness meditation. Brain Behav. 2014;4(3):337–47; and Lutz A, Brefczynski-Lewis J, Johnstone T, Davidson RJ. Regulation of the neural circuitry of emotion by compassion meditation: effects of meditative expertise. PLoS One. 2008;3(3):e1897.

6 Kang Y, Gray JR, Dovidio JF. The nondiscriminating heart: loving-kindness meditation training decreases implicit intergroup bias. J Exp Psychol Gen. 2014;143(3):1306–13.

7 Leung MK, Chan CC, Yin J, Lee CF, So KF, Lee TM. Increased gray matter volume in the right angular and posterior parahippocampal gyri in loving-kindness meditators. Soc Cogn Affect Neurosci. 2013;8(1):34–9.

8 Worthington EL, Berry JW, Hook JN, Davis DE, Scherer M, Griffin BJ, Wade NG, Yarhouse M, Ripley JS, Miller AJ, Sharp CB, Canter DE, Campana KL. Forgiveness-reconciliation and communication-conflict-resolution interventions versus retested controls in early married couples. J Couns Psychol. 2015;62(1):14–27.

9 Ricciardi E, Rota G, Sani L, Gentili C, Gaglianese A, Guazzelli M, Pietrini P. How the brain heals emotional wounds: the functional neuroanatomy of forgiveness. Front Hum Neurosci. 2013;7:839 doi:10.3389/fnhum.2013.00839.

10 Thompson LY, Snyder CR, Hoffman L, Michael ST, Rasmussen HN, Billings LS, Heinze L, Neufeld JE, Shorey HS, Roberts JC, Roberts DE. Dispositional forgiveness of self, others, and situations. J Pers. 2005;73(2):313–59.

11 Lawler KA, Younger JW, Piferi RL, Jobe RL, Edmondson KA, Jones WH. The unique effects of forgiveness on health: an exploration of pathways. J Behav Med. 2005;28(2):157–67.

12 Bono G, McCullough ME, Root LM. Forgiveness, feeling connected to others, and well-being: two longitudinal studies. Pers Soc Psychol Bull. 2008;34(2):182–95.

13 Fehr R, Gelfand MJ, Nag M. The road to forgiveness: a meta-analytic synthesis of its situational and dispositional correlates. Psychol Bull. 2010;136(5):894–914.

14 R. Heuer, "Psychology of Intelligence Analysis." Published by the U.S. Government, Center for the Study of Intelligence, Central Intelligence Agency, 1999.

15 Harris S, Sheth SA, Cohen MS. Functional neuroimaging of belief, disbelief, and uncertainty. Ann Neurol. 2008 Feb;63(2):141–7.

16 Seidel EM, Pfabigan DM, Hahn A, Sladky R, Grahl A, Paul K, Kraus C, Küblböck M, Kranz GS, Hummer A, Lanzenberger R, Windischberger C, Lamm C. Uncertainty during pain anticipation: the adaptive value of preparatory processes. Hum Brain Mapp. 2015;36(2):744–55.

17 Smith DF. Functional salutogenic mechanisms of the brain. Perspect Biol Med. 2002;45(3):319-28.

18 Tsujii T, Masuda S, Akiyama T, Watanabe S. The role of inferior frontal cortex in belief-bias reasoning: an rTMS study. Neuropsychologia. 2010;48(7):2005–8.

Part 3. Moving Toward Enlightenment

1 This version is a composite interpretation of a series of Rumi poems originally translated by A. G. Farhadi and Ibraham Gamard in an unpublished manuscript. Copyright 2014 by Mark Robert Waldman.

Chapter 10. Preparing for Enlightenment

1 Schippers MB, Roebroeck A, Renken R, Nanetti L, Keysers C. Mapping the information flow from one brain to another during gestural communication. Proc Natl Acad Sci USA. 2010;107(20):9388–93; and Schippers MB, Gazzola V, Goebel R, Keysers C. Playing charades in the fMRI: are mirror and/or mentalizing areas involved in gestural communication? PLoS One. 2009;4(8):e6801.

2 Csikszentmihalyi, Mihaly, *Flow: The Psychology of Optimal Experience* (New York: Harper and Row, 1990).

3 Ulrich M, Keller J, Hoenig K, Waller C, Grön G. Neural correlates of experimentally induced flow experiences. Neuroimage. 2014;86:194–202.

4 Barutta J, Gleichgerrcht E, Cornejo C, Ibáñez A. Neurodynamics of mind: the arrow illusion of conscious intentionality as downward causation. Integr Psychol Behav Sci. 2010;44(2):127–43; and Lloyd D. Functional MRI and the study of human consciousness. J Cogn Neurosci. 2002;14(6):818–31.

5 Neill J. Transcendence and transformation in the life patterns of women living with rheumatoid arthritis. ANS Adv Nurs Sci. 2002;24(4):27–47; and Wade GH. A concept analysis of personal transformation. J Adv Nurs. 1998;28(4):713–9.

6 Ullrich PM, Lutgendorf SK. Journaling about stressful events: effects of cognitive processing and emotional expression. Ann Behav Med. 2002;24(3):244–50.

7 Smith S, Anderson-Hanley C, Langrock A, Compas B. The effects of journaling for women with newly diagnosed breast cancer. Psychooncology. 2005;14(12):1075–82.

8 Schwartz RM, Reynolds CF, Thase ME, Frank E, Fasiczka AL, Haaga DAF. Optimal and normal affect balance in psychotherapy of major depression: evaluation of the balanced states of mind model. Behav Cogn Psychother. 2002 Oct; 30(4):439–450.

9 B. Fredrickson, *Positivity* (New York: Three Rivers Press, 2009).

10 Losada M, Heaphy E. The role of positivity and connectivity in the performance of business teams: a nonlinear dynamics model. Am Behav Scientist. 2004 47 (6):740–65.

11 J. Gottman, *What Predicts Divorce?: The Relationship Between Marital Processes and Marital Outcomes* (Hillsdale, NJ: Lawrence Erlbaum Associates, 1994).

12 Lundqvist LO, Zetterlund C, Richter HO. Effects of Feldenkrais method on chronic neck/scapular pain in people with visual impairment: a randomized controlled trial with one-year follow-up. Arch Phys Med Rehabil. 2014;95(9):1656–61.

13 Gallup AC, Eldakar OT. The thermoregulatory theory of yawning: what we know from over 5 years of research. Front Neurosci. 2013;6:188.

14 Gallup AC, Gallup GG. Yawning as a brain cooling mechanism: nasal breathing and forehead cooling diminish the incidence of contagious yawning. Evol Psychol. 2007;5:92–101.

15 Milbury K, Chaoul A, Biegler K, Wangyal T, Spelman A, Meyers CA, Arun B, Palmer JL, Taylor J, Cohen L. Tibetan sound meditation for cognitive dysfunction: results of a randomized controlled pilot trial. Psychooncology. 201322(10):2354–63

16 Sauvage C, Jissendi P, Seignan S, Manto M, Habas C. Brain areas involved in the control of speed during a motor sequence of the foot: real movement versus mental imagery. J Neuroradiol. 2013;40(4):267–80.

17 Kronk CM. Private speech in adolescents. Adolescence. 1994;29(116):781–804.

18 Tullett AM, Inzlicht M. The voice of self-control: blocking the inner voice increases impulsive responding. Acta Psychol (Amst). 2010;135(2):252–6.

19 Geva S, Jones PS, Crinion JT, Price CJ, Baron JC, Warburton EA. The neural correlates of inner speech defined by voxel-based lesion-symptom mapping. Brain. 2011;134(Pt 10):3071–82.

20 Johnstone T, van Reekum CM, Urry HL, Kalin NH, Davidson RJ. Failure to regulate: counterproductive recruitment of top-down prefrontal-subcortical circuitry in major depression. J Neurosci. 2007 Aug 15;27(33):8877–84.

21 Marchand WR. Neural mechanisms of mindfulness and meditation: evidence from neuroimaging studies. World J Radiol. 2014;6(7):471–9.

22 Bluett EJ, Homan KJ, Morrison KL, Levin ME, Twohig MP. Acceptance and commitment therapy for anxiety and OCD spectrum disorders: an empirical review. J Anxiety Disord. 2014;28(6):612–24.

23 Goyal M, Singh S, Sibinga EM, Gould NF, Rowland-Seymour A, Sharma R, Berger Z, Sleicher D, Maron DD, Shihab HM, Ranasinghe PD, Linn S, Saha S, Bass EB, Haythornthwaite JA. Meditation programs for psychological stress and well-being: a systematic review and meta-analysis. JAMA Intern Med. 2014;174(3):357–68.

24 Kieviet-Stijnen A, Visser A, Garssen B, Hudig W. Mindfulness-based stress reduction training for oncology patients: patients' appraisal and changes in well-being. Patient Educ Couns. 2008;72(3):436–42.

Chapter 11. Intensifying the Experience

1 Salimpoor VN, Benovoy M, Larcher K, Dagher A, Zatorre RJ. Anatomically distinct dopamine release during anticipation and experience of peak emotion to music. Nat Neurosci. 2011;14(2):257–62.

2 Nordbø RH, Gulliksen KS, Espeset EM, Skårderud F, Geller J, Holte A. Expanding the concept of motivation to change: the content of patients' wish to recover from anorexia nervosa. Int J Eat Disord. 2008;41(7):635–42.

3 Holland AE, Hill CJ, Jones AY, McDonald CF. Breathing exercises for chronic obstructive pulmonary disease. Cochrane Database Syst Rev. 2012;10:CD008250; and Cramer H, Krucoff C, Dobos G. Adverse events associated with yoga: a systematic review of published case reports and case series. PLoS One. 2013;8(10):e75515.

4 Pagnoni G, Cekic M, Guo Y. "Thinking about not-thinking": neural correlates of conceptual processing during Zen meditation. PLoS One. 2008;3(9):e3083.

5 Grant JA, Courtemanche J, Rainville P. A non-elaborative mental stance and decoupling of executive and pain-related cortices predicts low pain sensitivity in Zen meditators. Pain. 2011 Jan;152(1):150–6.

6 Isaacowitz DM. The gaze of the optimist. Pers Soc Psychol Bull. 2005 Mar;31(3):407–15.

7 Hurlburt RT, Heavey CL, Kelsey JM. Toward a phenomenology of inner speaking. Conscious Cogn. 2013;22(4):1477–94.

8 Cwik AJ. Associative dreaming: reverie and active imagination. J Anal Psychol. 2011;56(1):14–36.

9 Kalyani BG, Venkatasubramanian G, Arasappa R, Rao NP, Kalmady SV, Behere RV, Rao H, Vasudev MK, Gangadhar BN. Neurohemo-dynamic correlates of "OM" chanting: A pilot functional magnetic resonance imaging study. Int J Yoga. 2011;4(1):3–6.

10 Khalsa DS, Amen D, Hanks C, Money N, Newberg A. Cerebral blood flow changes during chanting meditation. Nucl Med Commun. 2009;30(12):956–61.

11 Gruzelier J. A theory of alpha/theta neurofeedback, creative performance enhancement, long distance functional connectivity and psychological integration. Cogn Process. 2009;10 Suppl 1:S101–9.

12 Sallustro F, Atwell CW. Body rocking, head banging, and head rolling in normal children. J Pediatr. 1978;93(4):704–8.

13 Wayne PM, Berkowitz DL, Litrownik DE, Buring JE, Yeh GY. What do we really know about the safety of Tai Chi? A systematic review of adverse event reports in randomized trials. Arch Phys Med Rehabil. 2014; 95(12):2470–83. Wayne PM, Walsh JN, Taylor-Piliae RE, Wells RE, Papp KV, Donovan NJ, Yeh GY. Effect of tai chi on cognitive performance in older adults: systematic review and meta-analysis. J Am Geriatr Soc. 2014;62(1):25–39.

14 R. Eisenberg, *The JPS Guide to Jewish Traditions* (Philadelphia: Jewish Publication Society, 2004).

15 Jones D. Social evolution: the ritual animal. Nature 2013 Jan 23. http://www.nature.com/news/social-evolution-the-ritual-animal-1.12256.

16 Panchin AY, Tuzhikov AI, Panchin YV. Midichlorians—the biomeme hypothesis: is there a microbial component to religious rituals? Biol Direct. 2014;9(1):14; and Umen J, Heitman J. Evolution of sex: mating rituals of a pre-metazoan. Curr Biol. 2013;23(22):R1006–8.

17 Frecska E, Luna LE. Neuro-ontological interpretation of spiritual experiences. Neuropsychopharmacol Hung. 2006;8(3):143–53.

18 Graybiel AM. Habits, rituals, and the evaluative brain. Annu Rev Neurosci. 2008;31:359–87.

19 Online Enclyclopedia Brittanica: http://www.britannica.com/EBchecked/topic/15965/Allah.

20 Tierney A, Dick F, Deutsch D, Sereno M. Speech versus song: multiple pitch-sensitive areas revealed by a naturally occurring musical illusion. Cereb Cortex. 2013;23(2):249–54.

21 Kokal I, Engel A, Kirschner S, Keysers C. Synchronized drumming enhances activity in the caudate and facilitates prosocial commitment—if the rhythm comes easily. PLoS One. 2011;6(11):e27272.

22 Boecker H, Sprenger T, Spilker ME, Henriksen G, Koppenhoefer M, Wagner KJ, Valet M, Berthele A, Tolle TR. The runner's high: opioidergic mechanisms in the human brain. Cereb Cortex. 2008 Nov;18(11):2523–31.

23 Masters KS. Hypnotic susceptibility, cognitive dissociation, and runner's high in a sample of marathon runners. Am J Clin Hypn. 1992 Jan;34(3):193–201.

24 Kokal I, Engel A, Kirschner S, Keysers C. Synchronized drumming enhances activity in the caudate and facilitates prosocial commitment—if the rhythm comes easily. PLoS One. 2011;6(11):e27272.

25 Bittman BB, Snyder C, Bruhn KT, Liebfreid F, Stevens CK, Westengard J, Umbach PO. Recreational music-making: an integrative group intervention for reducing burnout and improving mood states in first year associate degree nursing students: insights and economic impact. Int J Nurs Educ Scholarsh. 2004;1:Article12.

26 Bittman BB, Berk LS, Felten DL, Westengard J, Simonton OC, Pappas J, Ninehouser M. Composite effects of group drumming music therapy on modulation of neuroendocrine-immune parameters in normal subjects. Altern Ther Health Med. 2001;7(1):38–47.

27 Remedios R, Logothetis NK, Kayser C. Monkey drumming reveals common networks for perceiving vocal and nonvocal communication sounds. Proc Natl Acad Sci U S A. 2009 Oct 20;106(42):18010–5.

28 Patel AD, Iversen JR, Bregman MR, Schulz I. Experimental evidence for synchronization to a musical beat in a nonhuman animal. Curr Biol. 2009 May 26;19(10):827–30.

29 Kotiaho JS, Alatalo RV, Mappes J, Parri S. Adaptive significance of synchronous chorusing in an acoustically signalling wolf spider. Proc Biol Sci. 2004;271(1550):1847–50.

30 Ho P, Tsao JC, Bloch L, Zeltzer LK. The impact of group drumming on social-emotional behavior in low-income children. Evid Based Complement Alternat Med. 2011:250708.

31 Winkelman M. Complementary therapy for addiction: "drumming out drugs". Am J Public Health. 2003 Apr;93(4):647–51.

32 Manning C, Waldman M, Lindsey W, Newberg A, Cotter-Lockard D. Personal inner values—a key to effective face-to-face business communication. J Executive Education. 2013;11(1)37–65.

33 Dusek JA, Otu HH, Wohlhueter AL, Bhasin M, Zerbini LF, Joseph MG, Benson H, Libermann TA. Genomic counter-stress changes induced by the relaxation response. PLoS One. 2008;3(7):e2576.

Appendix. Tools and Resources to Foster Enlightenment

1 Larouche M, Côté G, Bélisle D, Lorrain D. Kind attention and non-judgment in mindfulness-based cognitive therapy applied to the treatment of insomnia: state of knowledge. Pathol Biol (Paris). 2014; 62(5):284–91.

2 Shennan C, Payne S, Fenlon D. What is the evidence for the use of mindfulness-based interventions in cancer care? A review. Psychooncology. 2011;20(7):681–97.

3 Matchim Y, Armer JM, Stewart BR. Mindfulness-based stress reduction among breast cancer survivors: a literature review and discussion. Oncol Nurs Forum. 2011;38(2):E61–71.

4 Williams D, Tricomi G, Gupta J, Janise A. Efficacy of burnout interventions in the medical education pipeline. Acad Psychiatry. 2014; 39(1):47–54.

5 Fortney L, Luchterhand C, Zakletskaia L, Zgierska A, Rakel D. Abbreviated mindfulness intervention for job satisfaction, quality of life, and compassion in primary care clinicians: a pilot study. Ann Fam Med. 2013;11(5):412–20.

6 Beddoe AE, Murphy SO. Does mindfulness decrease stress and foster empathy among nursing students? J Nurs Educ. 2004;43(7):305–12.

INDEX

Page numbers in **bold** indicate tables, *italics* indicate illustrations, and those followed by "n" indicate notes.

ABOUT THE AUTHORS

Andrew Newberg, MD, and Mark Waldman are the world's leading experts on spirituality and the brain. They have coauthored four books including *Words Can Change Your Brain: 12 Conversation Strategies to Build Trust, Resolve Conflict, and Increase Intimacy* and the national bestseller *How God Changes Your Brain*, picked by Oprah as a "must read" book in 2012. Together they have authored and edited over twenty books and over one hundred academic papers. Their research has been featured in *Time*, *Newsweek*, the *Washington Post*, the *New York Times*, *Forbes*, *Entrepreneur*, *Oprah Magazine*, and many others. They have both appeared on hundreds of radio and television programs, including PBS and NPR Radio.

Dr. Newberg is director of research at the Myrna Brind Center of Integrative Medicine at Thomas Jefferson University. He is also a professor in the Departments of Emergency Medicine and Radiology at Thomas Jefferson University. He is coauthor of the national bestseller *Why God Won't Go Away: Brain Science and the Biology of Belief* and several academic books including *Principles of Neurotheology.* If you'd like more information on Dr. Newberg, visit his website at www.AndrewNewberg.com.

Mark Waldman is executive MBA faculty at Loyola Marymount University. He also teaches at Holmes Institute, a theological seminary, and is a business and personal development NeuroCoach, integrating brain-based strategies drawn from his research on mindfulness, positivity, and cognitive training. He received the Distinguished Speaker award from the Mind Science Foundation. If you'd like more information on Mark and his speaking itinerary, visit his website at www.MarkRobertWaldman.com.